RELATING TO THINGS

RELATING TO THINGS

Design, Technology and the Artificial

Edited by Heather Wiltse

BLOOMSBURY VISUAL ARTS
LONDON • NEW YORK • OXFORD • NEW DELHI • SYDNEY

BLOOMSBURY VISUAL ARTS
Bloomsbury Publishing Plc
50 Bedford Square, London, WC1B 3DP, UK
1385 Broadway, New York, NY 10018, USA

BLOOMSBURY, BLOOMSBURY VISUAL ARTS and the Diana logo
are trademarks of Bloomsbury Publishing Plc

First published in Great Britain 2020

© Editorial content and introductions, Heather Wiltse, 2020
© Individual chapters, their authors, 2020

Heather Wiltse has asserted her right under the Copyright, Designs and
Patents Act, 1988, to be identified as Editor of this work.

For legal purposes the Acknowledgments on p. xiv constitute an
extension of this copyright page.

Cover designer: Marije de Haas

All rights reserved. No part of this publication may be reproduced or transmitted
in any form or by any means, electronic or mechanical, including photocopying,
recording, or any information storage or retrieval system, without prior
permission in writing from the publishers.

Bloomsbury Publishing Plc does not have any control over, or responsibility for,
any third-party websites referred to or in this book. All internet addresses given in
this book were correct at the time of going to press. The author and publisher regret
any inconvenience caused if addresses have changed or sites have ceased
to exist, but can accept no responsibility for any such changes.

A catalogue record for this book is available from the British Library.

A catalog record for this book is available from the Library of Congress.

ISBN: HB: 978-1-3501-2425-7
ePDF: 978-1-3501-2427-1
ePub: 978-1-3501-2426-4

Typeset by Integra Software Services Pvt. Ltd.
Printed and bound in Great Britain

To find out more about our authors and books visit www.bloomsbury.com
and sign up for our newsletters.

CONTENTS

List of Figures vii
List of Contributors xi
Acknowledgments xiv

Introduction: Relating to Things That Relate to Us
   Heather Wiltse 1

PART ONE CARING FOR THINGS THAT CARE FOR US 13

1. Privacy as Care in the Internet of Things
 D.E. Wittkower 15
2. Attachment to Things, Artifacts, Devices, Commodities: An Inconvenient Ethics of the Ordinary *Michel Puech* 31
3. The New Assisted Living: Caring for Alexa Caring for Us *Diane P. Michelfelder* 43

PART TWO LEARNING FROM THINGS THAT LEARN FROM US 59

4. Handling Things That Handle Us: Things Get to Know Who We Are and Tie Us Down to Who We Were *Bruno Gransche* 61
5. Can Ethics Be Learned? Video Games as an Ethical Sandbox *Fanny Verrax* 81

6   Casting Things as Partners in Design: Toward a More-than-Human Design Practice *Elisa Giaccardi* 99

PART THREE CONTROLLING THINGS THAT CONTROL US 133

7   Hostile Design and the Materiality of Surveillance
     Robert Rosenberger 135

8   A Tool for the Impact and Ethics of Technology: The Case of Interactive Screens in Public Spaces
     Steven Dorrestijn 151

9   Postphenomenology of Augmented Reality
     Galit Wellner 173

PART FOUR REVEALING THINGS THAT REVEAL US 189

10   Imagining Things: Unfolding the "of" in Philosophy of Technology, through Object-Oriented Ontology
     Yoni Van Den Eede 191

11   The Disappearing Acts of the Morse Things: A Design Inquiry into the Withdrawal of Things
     Ron Wakkary, Sabrina Hauser, and Doenja Oogjes 215

12   Revealing Relations of Fluid Assemblages
     Heather Wiltse 239

13   Designing Networks That Reveal Themselves
     Holly Robbins 255

14   Reflection and Commentary
     Erik Stolterman 271

Index 278

FIGURES

1.1 Realms of privacy as overlapping ranges of effective executive autonomy within an interdependent relationship 21

4.1 Three heuristic levels of autonomy/automation in human-technology relations 65

6.1 Portrait photos of participating scooters and scooterists. Photos by Wen-Wei Chang 106

6.2 Scooter portraits based on the transcripts of the "interview with things" conducted with professional actors. Image courtesy of Wen-Wei Chang 109

6.3 Pipe heater is the speculative concept for an open-ended device that reuses the heat produced during a ride for personally and socially meaningful activities (e.g., warming up food, sharing a hot drink). Photos by Wen-Wei Chang 110

6.4 Sound generator is the speculative concept of a smart audio component for a scooter's engine that is personalized according to one's riding patterns and needs for social expression. Photos by Wen-Wei Chang 111

6.5 Red light pointer and atmosphere meter are speculative concepts of smart dashboard components designed to bring people physically closer and create intimacy. Photos by Wen-Wei Chang 111

6.6 MakeDo is a speculative platform for DIY recipes where data collected from things are an integral part of the making process with the goal to foster creative

	dialogues between makers and things in democratized manufacturing. Image by Tal Amram 114
6.7	Example of the executed DIY recipe of a stool with sensing knots from the MakeDo community. Photos by Tal Amram 115
6.8	Resourceful Ageing: Selecting nonhuman participants via a combination of sensitization techniques and ethnographic fieldwork. Photo by Iohanna Nicenboim 117
6.9	Resourceful Ageing: Participating magnet, central to the resourcefulness of one of the human participants. Photo by Iohanna Nicenboim 118
6.10	Resourceful Ageing: Analysis of raw temporal events concerning co-usage of instrumented objects (i.e., relations among nonhuman participants). Data visualization by Yanxia Zhang 118
6.11	Resourceful Ageing: Visualization of machine learning interpretation of the co-usage of objects, from high to low probability of occurrence. Data visualization by Philips Design 119
6.12	Resourceful Ageing: *Connected Resources* is a family of sensors and actuators and an online service for adding digital capabilities to older people's everyday strategies of resourcefulness and empowering them in their relation with care technology. Images by Masako Kitazaki 122
6.13	Resourceful Ageing: Once in use, *Connected Resources* learn from the way in which they are combined and deployed 122
6.14	Resourceful Ageing: Scenario of a resourceful arrangement created by an older woman waiting for a delivery and with a mild hearing impairment, where one object visibly lights up when another remote object detects sound. Movie by Andreas D'Hollandere 123

7.1	Subway bench, New York City, United States (photo by author)	136
7.2	Ledge spikes, San Francisco, United States (photo by author)	137
7.3	Tube platform security camera, London, England (photo by author)	141
7.4	Outdoor security camera mounted on a pole and protected by pointed fencing, Winchester, England (photo by author)	141
7.5	Parking lot signage in Boston, United States (photo by author)	146
8.1	Product Impact Tool Model	153
8.2	One example of product impact from the online Product Impact Tool	153
8.3	Project name display on a screen on a pilot day. Source: http://www.actmedialab.nl/tweede-pilot-observe/	155
8.4	People in view of an interactive screen system. Source: http://www.actmedialab.nl/tweede-pilot-observe/	155
11.1	A set of Morse Things	216
11.2	Participant photos of Morse Things in their homes	223
11.3	Each Morse Things set included a large bowl, medium bowl, a cup plus a Wi-Fi hub and a set of instructions	224
11.4	Hannah's concept of Morse Things as part of daily routines	227
11.5	Olivia imagined the Morse Things could hack into other things to join their network	228
11.6	Ella's concept for finding and containing things	229
11.7	Spencer's concept of having the Morse Things as a Wi-Fi repeater and watering system	231
13.1	Thingformation, an IoT care labeling system for product packaging. Design and image: Beyond/IO	261

13.2 The Transparent Charging Station is an electric car charging station that allows people to negotiate how much of their battery is to be charged and within what time period according to the networked constraints of an electric energy grid. Design and image: The Incredible Machine 263

13.3 The interface of the Transparent Charging Station demonstrates the constraints and demands on the electric grid with a Tetris-like screen. The hourglass-like figure in the middle illustrates what energy is predicted to be available on the grid over the course of the day. This station has three different charging ports, each of which is controlled by two dials along the bottom. In turning the dials, a driver negotiates how much of her battery is to be changed and in what timeframe within the constraints of the energy predicted to be available on the grid. As the driver turns these dials, the screen also illustrates how their request impacts those made by the others charging at this port 264

CONTRIBUTORS

Steven Dorrestijn is Head of the research group Ethics & Technology at Saxion University of Applied Sciences, the Netherlands. In 2012 Dorrestijn completed his PhD thesis at the University of Twente, the Netherlands (*The design of our own lives: Technical mediation and subjectivation after Michel Foucault*). Previously he studied philosophy in Paris and philosophy and mechanical engineering in Twente. Dorrestijn's research and publications focus on the philosophy and ethics of technology, the work of Michel Foucault applied to technology, and the integration into design of knowledge about the impact of technology (Product Impact Tool). His website: www.stevendorrestijn.nl.

Elisa Giaccardi is Chair of Interactive Media Design at Delft University of Technology, the Netherlands, and Guest Professor of Post-Industrial Design at Umeå Institute of Design. After groundbreaking work in metadesign, collaborative and open design processes, Giaccardi has during the last years focused on the challenges that a permeating digitalization means for the field of design.

Bruno Gransche has been a philosopher at the Institute of Advanced Studies (FoKoS) at the University of Siegen since 2017. He works in the fields of philosophy of technology, ethics, and future-oriented thinking. He is a research fellow at the Fraunhofer ISI in Karlsruhe, where he worked as a philosopher and Foresight expert until 2016.

Sabrina Hauser is Postdoctoral Research Fellow at the Umeå Institute of Design. She received her PhD in 2018 from the School of Interactive Arts and Technology at Simon Fraser University. Sabrina holds a Masters in Design from Hochschule für Gestaltung Schwäbisch Gmünd and a Diplom (BSc Hons.) in Information Science from Hochschule Darmstadt. Situated at the intersections between interaction design, human-computer-interaction, and philosophy of technology, her research often looks at human-technology relations and technological mediation through approaches like design ethnographies, research through design, and speculative design. Her website: www.sabrinahauser.com.

Diane P. Michelfelder is Professor of philosophy at Macalester College in the United States. Her research interests focus on the philosophy of technology and engineering. A former president of the Society for Philosophy and Technology, her most recent book, edited with Byron Newberry and Qin Zhu, is *Philosophy and Engineering: Exploring Boundaries, Expanding Connections* (Springer 2016).

Doenja Oogjes is a PhD student whose research explores the design of digital domestic technologies using speculative design and integrating indirect ways technologies mediate our everyday. She holds a Bachelors and a Masters in Industrial Design from the Technical University of Eindhoven, Netherlands.

Michel Puech's academic background is in classical European philosophy. He has published books and articles on the philosophy of technology, the concept of the "sustainable," and more broadly on new value systems. His current work focuses on the notion of *wisdom*. His recent book is *The Ethics of Ordinary Technology* (2016).

Holly Robbins is a postdoctoral researcher at Eindhoven University of Technology. Her work blends industrial design with philosophy of technology, examining and experimenting with how design can make complex, networked, and data-intensive technologies legible and accessible for lay people. With a collection of professional designers, she co-founded the Just Things Foundation to promote and advocate for the responsible design of technologies.

Robert Rosenberger is Associate Professor in the School of Public Policy at the Georgia Institute of Technology. His edited books include *Postphenomenological Investigations: Essays on Human-Technology Relations* (coedited with P.P. Verbeek) and the interview book *Philosophy of Science: 5 Questions*. His polemical mini-monograph is entitled *Callous Objects: Designs Against the Homeless*.

Erik Stolterman is Professor in informatics and Senior Executive Associate Dean of the School of Informatics, Computing, and Engineering at Indiana University, Bloomington. Stolterman's main work is within HCI, interfaces, interactivity, interaction design, design practice, philosophy, and theory of design. His latest book is *Things That Keep Us Busy—The Elements of Interaction* (MIT Press, 2017).

Yoni Van Den Eede is Lecturer and Researcher in Philosophy, affiliated with the Centre for Ethics and Humanism, at the Free University of Brussels (Vrije Universiteit Brussel), Belgium. He is among others the author of *The Beauty of Detours* (2019) and *Amor Technologiae* (2012), and the coeditor of *Postphenomenology and Media* (2017).

Fanny Verrax works as a consultant for various organizations after a decade in academia, exploring science and technology studies, environmental philosophy, and professional ethics. In this chapter, she combines her interest in ethics as a researcher and an educator with her passion for a typically underrated human experience: videogames.

Ron Wakkary is Professor in the School of Interactive Arts and Technology (SIAT) at Simon Fraser University (SFU), where he established the Everyday Design Studio. He is also a Design United Visiting Professor and Chair of the Impact of Interaction Design on Everyday Life in the Industrial Design Department at Eindhoven University of Technology (TU/e).

Galit Wellner is Senior Lecturer at the NB School of Design, Israel. She also teaches philosophy of technology at Tel Aviv University and Bezalel. Her research focuses on digital postphenomenology. She published several peer-reviewed articles, book chapters, and books including *A Postphenomenological Inquiry of Cellphones* (2015: Lexington Books).

Heather Wiltse is Associate Professor at Umeå Institute of Design, Umeå University (Sweden). Her transdisciplinary research is concerned with the roles and consequences of connected, responsive, changing things and their design. Her recent book (coauthored with Johan Redström) is *Changing Things: The Future of Objects in a Digital World* (2019).

D.E. Wittkower is Associate Professor of philosophy, Old Dominion University. He teaches on philosophy of technology, philosophy of social media, and information literacy and digital culture. In addition to six books on philosophy for a general audience, he is author or coauthor of forty-one book chapters and journal articles.

ACKNOWLEDGMENTS

This project more or less began in the fall of 2016, when Michel Puech contacted me with the idea of proposing a session for the SPT 2017 conference in Darmstadt related to the theme of my research as it intersected with that of the conference on the "grammar of things." That eventually resulted in a successful double panel, and it was after SPT that Michel again got in touch to promote the idea of expanding it into an edited book project. We began that work as coeditors, but circumstances unfortunately prevented Michel from continuing in that role. However, it remains the case that this book would not exist without his initiative.

Much of my work in the last few years that I build on here has been done in collaboration with Johan Redström, who also generously served as a sounding board throughout this project. Erik Stolterman agreed to join for the book phase of the project and provided valuable advice about how to manage it and also write the funding and book proposals. Thanks also to Clive Dilnot for his advice and support for the book.

The workshop meeting that almost all of the authors had together in Umeå in the north of Sweden in the spring of 2018 was crucial for helping this book to develop in the way that it did, and that was enabled by awarded research initiation funding from Riksbankens Jubileumsfond. Maria Göransdotter and Corné de Beer at Umeå Institute of Design provided crucial support with ironing out the budget details for that application. My own continuing work on the project has been enabled by awarded research project funding from the Marianne and Marcus Wallenberg Foundation.

A big part of my inspiration for the way that I set up the collaborative process around this book was provided by Tarleton Gillespie, Pablo J. Boczkowski, and Kirsten A. Foot, in their introduction to the 2014 edited book *Media Technologies: Essays on Communication, Materiality, and Society*. From this text, I understood the importance of intensive in-person collaboration at a workshop, with the objective of building a shared inquiry rather than a collection of individual contributions. Their project provided a model that fueled my tenacity in aiming for those objectives, even as it meant asking a lot of the authors.

My thanks to Claire Collins at Bloomsbury for her expert management of the publication process and for her patience whenever I asked, again, if there were any updates yet.

Finally, deepest thanks to the fantastic authors who agreed to be part of this project and whose positive energy, enthusiasm, and goodwill made it the kind of academic dream that one can scarcely believe does, sometimes, actually come true.

INTRODUCTION: RELATING TO THINGS THAT RELATE TO US

Heather Wiltse

The texture of everyday life has always been woven in relation to things. From a child's beloved blanket to a favorite toy to a first bicycle or car to the many other large and small artifacts that serve our needs and desires as humans, we have always related to things.

Yet the ways in which things relate to us seem to be becoming ever more active and assertive. No longer merely inert matter that we must actively animate and press into the service of our needs, they are perhaps more like the butler Jeeves, the iconic character created by P.G. Wodehouse in his books: ever ready to attentively and discreetly provide whatever practical and emotional comforts are perceived as needed to smooth the carrying out of everyday activities or to soothe a rumpled soul. It is almost as if we can all have a Jeeves, or actually many, in our pockets, homes, and environments. Having a rough Monday? There is a recommended playlist for that. Need to get up at a certain time, but want the process to be as gentle as possible? There is an app for that to monitor your sleep cycle and wake you at the optimal moment. Need a reminder to get to your next appointment on time? Your smart assistant is on it. Should you really be moving a bit more? Your smart watch coach can always be with you, ready to prod and praise as needed.

Or at least this is the promise (or perhaps premise). The reality, however, is rather more complex. As with all kinds of relations and relationships, there can be misunderstandings, glitches, and breakdowns that call for ongoing repair work. The qualities of relations are also important; not so much static relations of the kind that might be drawn on some kind of network diagram, these are rather ongoing and dynamic interactions that evolve over time. And, indeed, these relations also have qualities of relationships: there can be significant relations with things that have history, durability, emotional engagement, mutual learning, ethical significance, and care. In fact, this longer-term involvement over time is becoming increasingly common, often framed in the language of things that are smart, responsive, adaptive, learning, and so on.

Relations and relationships can fall somewhere on a spectrum between diminishing and enriching, enabling us to become entrenched in destructive habits or calling us into becoming better versions of ourselves. They can exist over various temporalities, from momentary interaction to periodic encounter or long-term engagement. The balance of agency, power and knowledge, give and take in relationships can be unequal and also shift over time. There can be relations of care and service but also relations of domination, manipulation, exploitation, and control.

In fact, it seems that many of the relations people now maintain with things could be described as dysfunctional. It is now common wisdom that "if you're not paying for it, you're the product." Being a product means being continually monitored and targeted as consumers, every possible situation and mood, a "need state," and every occasion, an "addressable moment."[1] Things address themselves to our physical, mental, and emotional states in sometimes quite intimate ways, in many cases made even more persistent by the need to persuade or control behavior (to make a purchase or limit insurance risk, perhaps) and find ever more effective ways to ensure that particular forms of advertising content land squarely on their targets. As an executive at the location tracking company GroundTruth told *The New York Times*: "We look to understand who a person is, based on where they've been and where they're going, in order to influence what they're going to do next" (Valentino-DeVries et al. 2018). This is done through the mediation of smartphones—things that many of us have with us at all times and trust to help with managing many aspects of everyday life (those devious little snitches).

Although many people are aware on some level that data about them and their activities is being collected and distributed far and wide, it can still come as a shock when confronted with the scope, scale, and precision of this industrial enterprise. But the reality is that we now live in an attention economy (Wu 2016) running on the logic of surveillance capitalism (Zuboff 2019) and platform capitalism (Srnicek 2017), in which data is the new basic resource that is processed to generate value (The Economist 2017a, 2017b). Things relate to us in ways that are driven by marketing logics: turning us into precisely segmented audiences, trying to create needs and shape behavior, and monitoring the success of these efforts with precision, on a large scale, and in realtime. They are often designed to be addictive and to make it difficult for users to opt out of data collection (i.e., basic resource generation) through "dark patterns" of interface design. They encode everyday activity—the raw material of surveillance capitalism—in computational form and render it comparable and computable by those that operate the platforms (Alaimo and Kallinikos 2017). These kinds of connected things (including apps, websites, and so on as well as more physical things) continue to proliferate: serving needs and (manufactured) desires and visions of "smart cities" and similar; becoming more responsive, more pervasive, and more precisely tuned to serve their functions— only some of which involve generating value for end users. Seen in this light, things can take on a rather sinister and gloomy cast; and understanding them and the relations we have with them becomes a matter of fighting to maintain (or regain) our capacities for self-determination and integrity, before it is too late. In Shoshana Zuboff's powerful phrasing, we must fight to regain our "right to the future tense" (Zuboff 2019).

And yet, these things and the ways in which they relate to us can also be wonderful, almost magical, as they open up new possibilities. Things can provide different perspectives on the world and our selves and serve as creative partners in the production of cultural forms, creation of knowledge, care for ourselves and others, and even the design of our own lives. The dark side of addiction, manipulation, exploitation, control, and so on, always has its inverse in this rich capacity to serve more positive virtues and the best human potential. These are always choices, ones out of the many that add up to the construction of our shared artificial world and practices in it: our largest collective design project as humans.

While it might be possible to identify relational dynamics in interactions with both human and non-human entities, they are most pronounced in the case of entities that actively and through their own agency also relate to us. Until relatively recently, that category consisted of humans and some animals. Now, networked computational things have the capacity to actively relate to others in their environments.[2] As ordinary things are made responsive and connected, and more and more complex computational processes and network connections are packaged as things available for interaction, the world is increasingly textured by non-human entities that actively relate to us.

Other evidence of this phenomenon might be that some contemporary things also seem to be more likely to have or be assigned names. Some share the same addressable moniker across many physical embodiments: Alexa, Siri, Google. This is in itself quite interesting, that it is possible to point to a physical thing and to interact with it in a certain context, but to also know that it is in fact at the same time in a sense part of one larger thing: a vast network aggregating all interactions (minus the few by those who have chosen and then managed to figure out how to opt out) and based on these "learning" how to more effectively relate to humans. Other things seem to call out to be named, the act that serves as a relational primitive when bringing a new member into the circle of family relations. After all, how can one not name a robotic vacuum cleaner that trundles along in one's home, bumping into things and attempting to chew power cables and shoe laces with the determination and enthusiasm of a puppy?[3]

These dynamics around new kinds of human-technology relations can be addressed from the perspective of technological development, but it is vitally important to complement this perspective with one attuned to the roles that these things play in human experience and affairs. If we are going to have new types of relations and relationships, we need corresponding new ways of understanding them, as well as practical strategies and tactics for managing them—perhaps even including interventions, therapy, and breakups when relations go awry in destructive ways.[4]

The importance and urgency of this enterprise cannot be overstated. There is growing awareness and media coverage around the massive collection of personal information and its privacy implications, as well as concern about negative consequences of spending extensive amounts of time interacting with and through devices. But the questions underlying these serious concerns are about much more than screen time and how different actors in society choose to use technological tools. They include what these things mean to and for us, how they relate to us, and how others can now relate to and access us through things that we use every day. There is a pressing need to move past raising awareness to examining consequences, and from criticizing existing arrangements to envisioning how they might be better configured and negotiated (or at least, as a first step, disrupted).

There has been a recent and welcome "thing turn" across various disciplines that is relevant and, in some ways, helpful for this enterprise. But how to go about turning to properly examine the contemporary things that now surround and relate to us in often opaque ways? Where are the best starting points and most productive angles that can help us get to the heart of the matter to clarify what is at stake, and what alternatives are possible?

This book is an attempt to find out. It works at the intersection of philosophy of technology and design, where it is possible to ask fundamental questions about the role of things in human affairs and also how they might be designed differently to serve more desirable forms of life. As a whole, the book is a collaborative philosophical inquiry into the nature and consequences of contemporary technological things; and it is a design inquiry into the character of the artificial, and possibilities for how things might be otherwise.

Background

Fortunately, there is much that we have to build on. While each chapter builds up its own particular background, there are some overarching themes that can be identified.

First, philosophy of technology helps with identifying, articulating, and getting to the bottom of important fundamental questions. It is broadly concerned with the role and character of technologies in human affairs and questions of how to act in relation to them. It addresses basic questions in connection to technology, such as: What does it mean to be human and live a good life (in relation to technologies)? What are (technological) things? How do we perceive and act in the world (through the mediation of technologies)? What is the human condition (in an artificial world), and how can and should we cope with it? How do and should we relate to (human and nonhuman) others?

One particular subfield of philosophy of technology that many of the contributors work with to varying degrees is that of postphenomenology. This is a tradition that grew out of a combination of phenomenology and pragmatism and that emphasized the ways in which technologies often mediate and shape the access that we as humans have to our worlds (Ihde 2008). Its focus on technological mediation and schema for mapping different kinds of human-technology-world relations have made it quite practically as well as philosophically useful in exploring particular empirical cases.

Another main touchpoint in the book is that of design. While there are many different specialized design disciplines, the common foundation that they share is an orientation toward creating intentional change in the world (Nelson and Stolterman 2012). The overall context in which design operates is one in which the artificial—in other words, the designed—now constitutes the horizon of our existence as humans (Dilnot 2015). Design research and theory explore the character of design and designed things; design processes, materials, and outcomes; design cultures and histories; design agencies and responsibilities; and, perhaps most of all, possible designed futures (all design is inherently about the future, after all).

In design research, theory provides conceptual tools to think with (Stolterman 2008), and designed things or prototypes provide definitions and theory about what a particular kind of design space can be like (Redström 2017). Theory in these contexts is propositional, exploratory, and generative. Its success is measured not by proximity to some universal truth or ideal but by what useful perspectives it opens up and what it can be used to create.

Both philosophy of technology and design theory are challenged by digital things, which operate according to different logics than even quite complicated analog things. Computation is central in the field of human-computer interaction (HCI), but this field has been traditionally more focused on usability and possibilities to develop applications for particular kinds of use cases than it has on more philosophical and critical questions about their larger consequences (albeit with some notable exceptions). Recent design-oriented material culture studies, while a welcome expansion and intersection in the "thing turn," does not quite account for the specificities of digital materials (beyond recognition of their physical embodiments and supporting physical infrastructures). Interdisciplinary and activist research projects in areas such as AI, data, algorithms, and their ethics are becoming more common (and are desperately needed).

Our project presented in this book works within this space of existing traditions of inquiry and recognition of the need for new directions, combinations, and trajectories. It addresses fundamental issues regarding what is involved and at stake in relations to and among these new kinds of digital networked things that have not yet been adequately theorized in terms of their roles and consequences in everyday life and society. So far, very few books in philosophy consider objects and artifacts, and few books in design address the role and character of things at the level of philosophical foundations and critical reflection. While the individual chapters are focused on more particular issues, taken as a combined whole, this book addresses these issues squarely.

Problematics and Process

The big overarching question, then, is: How to relate to these things that relate to us? How do we[5] relate to them in the everyday, negotiating the complexities of life and the social world as they try to help but inevitably at times break down and require relational reparation? Or how and at what point do relations with things become dysfunctional or out of healthy and productive balance? What do these relations bring to our lives and sense of self? What, if anything, do we give in return? How, where, and by whom are the terms of these multiple and often nested relations negotiated and settled? And how can we begin to make sense of these things that relate to us and to each other in ways that are often active and illegible, offering interesting and exciting possibilities but also in some cases concealing what they really do and how they may even use or control us?

These questions constitute the core problematics explored in this book. An earlier version of these served as the initial provocation and invitation to some of the authors here to join what became a double panel session at the SPT 2017 conference (the conference of the Society for Philosophy and Technology) in Darmstadt, Germany, in the summer of 2017. Given the success and interest generated by this panel, not least among those of us who participated, we decided to expand it into a book project (again, as with the panel, prompted by initiative and enthusiasm from Michel Puech). For this we invited others who presented on similar themes at the conference to join us, as well as a few design researchers doing work in this space.

Awarded research initiation funding from Riksbankens Jubileumsfond in Sweden allowed us to meet together as this expanded group for two days in Umeå in the spring of 2018. The purpose of this meeting was not only to enable collective work on these problematics but also more generally to provide a space and structure for exchange among participants (especially those based in different fields who do not normally have venues for interaction) in order to open up for further connection and collaboration.

The project was founded on resolute methodological openness, with the goal of doing philosophy and making design theory that is capable of accounting for what is going on with these things that relate to us, and in ways that can enable incisive critique and (design) intervention. This requires analytic and conceptual work in both philosophy of technology and design, and combining the tools and sensitivities that both bring.

At the workshop in Umeå, the bulk of the time was spent on working together to identify common themes in the individual contributions that could serve to further clarify core foundational issues in this larger problem space. The themes eventually arrived at after significant time, effort, and energetic discussion are reflected in the parts of the book: *Caring for things that care for us*, *Learning from things that learn from us*, *Controlling things that control us*, and *Revealing things that reveal us*.

Clearly, the overarching formulation of *relating to things that relate to us* was generative enough to warrant further derivation, pointing to new, and newly experienced, forms of reciprocity in relations between things and us. This is not to say that the reciprocity is equivalent: the metaphor goes only so far (as Bruno Gransche nicely highlights in his contribution). Indeed, a lingering concern and shortcoming of these somewhat pithy formulations is that the suggested parallelism of human and thing relations obscures very real differences that remain. This was one of the sticking points in developing the themes, as we wrestled with whether we could really say that our contributions addressed both human and thing caring/learning/controlling/revealing. However, at the same time, interesting insights began to emerge in thinking this through in these terms.

While it might be rather common to see even ordinary things as *caring* for us and our needs, and more or less in ways for which they were designed, the idea of us also caring for these things as they care for us opens up new possibilities for both pragmatic interaction and design possibilities, and ethical responsibility. *Learning* points to the fact that things often now (machine) learn from us how to process the world, as we may also learn from them and the new perspectives that they can open up. *Controlling* is perhaps the most uncomfortable theme, since out-of-control technologies have been a persistent human fear. The chapters here explore how things can be used as tools to control others and also exert subtler forms of governance over perception and behavior. The question of how to exert control over technology that controls or governs us assumes fresh shape and urgency in this context. Finally, *revealing* gets at the basic difficulties of revealing what things are and do, both in terms of the specifics of contemporary connected things but also, more fundamentally, things in general. Moreover, things that relate to us are often animated by the purpose of revealing human activity rendered in the form of data, the new basic resource fueling capital accumulation—while concealing the fact that they do so. Revealing these relations then becomes a countermove of resistance and a step toward envisioning alternatives.

Thus, even with a few concerns and reservations, these themes stuck. The theme groups that were eventually formed in Umeå were tasked with together writing an overview of their theme, and the results appear here as the introductions to each part of the book.

These themes are an initial map of the problem space and the design space of things that relate to us. They start to point to what is at stake in relations with things, as explored in more detail in each of the chapters.

It should be noted that the perspectives we develop here are inevitably partial. While the composition of the group involved in the project was very intentional, it is also in some ways a historical accident and snapshot that is reflected in the book (which does not mean this snapshot is not valuable). Although the most common theoretical orientations engaged here are mainly in philosophy of technology, design research, and design-oriented human-computer interaction (HCI), other kinds of critical and design theory are also of general relevance. The possibility to bring insights and tools from these other discourses to bear in considering relations with contemporary things provides an opening for future work that could complement what we do in this book.

Overview of the Book

The book begins with *care*, and Dylan Wittkower's care-theoretic critique of the privacy paradox using cases from the Internet of Things (IoT). He here uses a phenomenology of privacy as it appears in interpersonal relationships to rethink privacy in digital contexts, arguing that in relations with caring things, people tend to draw on intuitions about how privacy works in caring human relationships. In this sense privacy is not so much control of information as a constant negotiation of intimacy. Next, in his contribution, Michel Puech makes a case for extending ethical consideration to things. Building on attachment theory in psychology, he sees ordinary things and objects as attachment-related artifacts that provide care in some sense; and building on wisdom ethics, he argues for new ways of engaging and valuing the objects that populate our technosphere and infosphere. Diane Michelfelder, in her chapter, considers relations with one particular thing: Amazon's tabletop digital assistant, Echo (more commonly known by the wake-up name Alexa). Observing in a discourse analysis of reviews of the product left on Amazon how readily people form attachment to Alexa (as seen in frequent use of the word "love"), she considers how caring interactions with it, specifically in the form of helping it to learn and train its algorithms, might help humans to build the capacity of caring for others. However, this type of empathetic relation raises serious privacy concerns, and she suggests possibilities for approaching the design of such assistants in ways that enable them to assist also with preserving privacy. The chapter concludes with a consideration of broader themes associated with digital assistants and care.

In Part II on *learning*, Bruno Gransche adds another riff on the main theme with his consideration of "handling things that handle us." He sees the rise in autonomous systems as marking a new stage in human-technology relations, one in which humans shift from craftsman to conductor. The chapter carefully builds up the argument that human learning of skills in service of particular goals is progressively displaced, as technologies increasingly provide what humans need or want without requiring much effort or skill on their part. Eventually, this seems to lead to a situation in which even the setting of goals is handled by intelligent systems that know how to serve and handle their "users." Moving into slightly more optimistic terrain, Fanny Verrax, in her contribution, outlines a way of seeing first-person video games as an ethical sandbox that can expand moral imagination in a safe learning environment. In this environment people are able to experience extreme power but also radical weakness, and to practice ethical virtues. Elisa Giaccardi, in her chapter, sees learning in the role that things can play in design. Current technological advancements such as in the Internet of Things, machine learning, and artificial intelligence mean that things can make things, too, raising fundamental issues for the role of designers in relation to data-enabled things. The chapter presents work that suggests that things, now able to sense and perform autonomously, can have access to perspectives and fields that we as humans do not. Through these capabilities, they can become partners in design, offering a thing perspective that can expand understanding and problematize what is taken for granted.

Beginning the part on *controlling*, Robert Rosenberger expands on previous work on hostile technology to consider how security camera surveillance also can constitute a form of control. Although they may not exert physical obstruction as in the case of spikes on ledges or armrests that make benches unsuited for sleeping, he explores how surveillance cameras in public space can also be considered a form of hostile design. In his chapter, Steven Dorrestijn looks at how to ethically relate to things that relate to us. He does this through using his Product Impact Tool to evaluate interactive screens in public space. This tool allows for unpacking different types of impact that technological things can have, making the connection between philosophical reflection and technology design. Another examination of technologies in public spaces is provided in the next chapter by Galit Wellner, but here in relation to augmented reality (AR) and from a postphenomenological perspective. Her analysis is structured in terms of the "immersiveness" of specific AR technologies and degree of necessity, from a leisure game (the phenomenally popular Pokémon Go) to, at the other end, a video speculation of an everyday reality in which AR is completely pervasive.

The final part on *revealing* begins with Yoni Van Den Eede's chapter probing presuppositions about things and relations at a quite fundamental level and then trying to reimagine them. This connects to the long-standing debate in philosophy on substance and relation, which is also present in philosophy of technology. Van Den Eede sees a possible way out through object-oriented ontology (OOO), which enables a blending of substance and relation and requires deep reconsideration of usual assumptions about objects and our relations to them. The following chapter by Ron Wakkary, Sabrina Hauser, and Doenja Oogjes returns to a somewhat more concrete level with their description of their Morse Things project, in which they used material speculation and co-speculation to come closer to the ways things withdraw from us. The project investigates what happens when everyday things are networked together and able to communicate with each other, or in other words, the IoT, through a thing-oriented approach. It highlights the ways in which these things withdraw from both those who live with them as well as from the design researchers who made them, as there is a gap in intelligibility of relations. Next, in my own chapter, I look at the multiple relations entailed in things that are *fluid assemblages*—fluid in that their forms and functions change across contexts and users, and assemblages in that they are emergent entities composed of a variety of interconnected and heterogeneous components. These multiple relations involve different actors and agencies that relate to and through things in different ways and determine the roles that things play in the world. The chapter concludes with a reflection on the need for new kinds of breakdown in order to understand these kinds of relations, and particularly their emerging role as key mediators of surveillance capitalism. Finally, in the concluding chapter, Holly Robbins tries to figure out how to design networks that reveal themselves. The things commonly referred to as "smart" technologies are powered by entire networks that exist behind the devices, both enabling the functionality of the devices we hold in our hands and use but also learning about us and dispersing that information throughout the network. In fact, she argues that we and these networks co-constitute each other. Designing to reveal these networks and make them more relatable is thus an important design challenge and one explored in this chapter through conceptual design work that treats these networks as materials to work with.

The book concludes with a commentary and reflection from Erik Stolterman, in which he reflects on why we even need to give serious consideration to things and our relations with them, and what it means to do that.

The chapters in this book and the compelling examples used show that things that relate to us can be intimately interwoven with our everyday interactions, sense of self, moral development, and more, while also possibly playing other roles and serving other relations and agendas. So, then: *How to relate to these things that relate to us?*

Notes

1 For one striking example of these dynamics, see Spotify's information for brands (https://spotifyforbrands.com/en-GB/).
2 Environments in the case of digital networked things are defined by network topology, which might overlap, but is not necessarily coextensive, with physical space.
3 The commonality of naming Roombas and other robotic vacuum cleaners can be seen through a quick web search, and Michelfelder also discusses research on relations with Roombas in her chapter. (The robotic vacuum cleaner in my home is called Robbie, a name revealed as disappointingly but perhaps unsurprisingly unoriginal by the aforementioned web search.)
4 One example of a therapeutic intervention is the addition of the "screen time" feature in Apple's devices, which provides a report of how much time people spend using the device and for which kinds of activities. In the best case, this might help to enable reflection and more intentional use.
5 The "we" here includes all who live with these technologies but also reflects the experiences that the chapter authors have with things that are often central to the reflections, observations, and analyses in their texts. There is no higher or external vantage point to be found: we are always already entangled in relations, and this is in fact essential for making sense of them.

References

Alaimo, Cristina and Jannis Kallinikos. 2017. "Computing the Everyday: Social Media as Data Platforms." *The Information Society* 4 (33): 175–191. https://dx.doi.org/10.1080/01972243.2017.1318327.

Dilnot, Clive. 2015. "History, Design, Futures: Contending with What We Have Made." In *Design and the Question of History*, edited by Tony Fry, Clive Dilnot, and Susan C. Stewart, 131–271. London: Bloomsbury.

The Economist. 2017a. "Data Is Giving Rise to a New Economy." *The Economist*, May 6, 2017. https://www.economist.com/news/briefing/21721634-how-it-shaping-up-datagiving-rise-new-economy.

The Economist. 2017b. "The World's Most Valuable Resource Is No Longer Oil, but Data," May 6, 2017. https://www.economist.com/news/leaders/21721656-data-economydemands-new-approach-antitrust-rules-worlds-most-valuable-resource.

Ihde, Don. 2008. "Introduction: Postphenomenological Research." *Human Studies* 31 (1): 1–9.

Nelson, Harold G. and Erik Stolterman. 2012. *The Design Way: Intentional Change in an Unpredictable World*. Cambridge, MA; London: MIT Press.

Redström, Johan. 2017. *Making Design Theory*. Cambridge, MA; London: MIT Press.

Srnicek, Nick. 2017. *Platform Capitalism*. Cambridge, UK, and Malden, MA, USA: Polity.

Stolterman, Erik. 2008. "The Nature of Design Practice and Implications for Interaction Design Research." *International Journal of Design* 2 (1): 55–65.

The Economist. 2017a. "Data Is Giving Rise to a New Economy." *The Economist*, May 6, 2017. https://www.economist.com/news/briefing/21721634-how-it-shaping-up-data-giving-rise-new-economy.

The Economist. 2017b. "The World's Most Valuable Resource Is No Longer Oil, but Data," May 6, 2017. https://www.economist.com/news/leaders/21721656-data-economy-demands-new-approach-antitrust-rules-worlds-most-valuable-resource.

Valentino-DeVries, Jennifer, Natasha Singer, Michael H. Keller, and Aaron Krolik. 2018. "Your Apps Know Where You Were Last Night, and They're Not Keeping It Secret." *The New York Times*, December 10, 2018. https://nyti.ms/2G6l9XC.

Wu, Tim. 2016. *The Attention Merchants: The Epic Scramble to Get Inside Our Heads*. New York: Knopf.

Zuboff, Shoshana. 2019. *The Age of Surveillance Capitalism: The Fight for a Human Future at the New Frontier of Power*. London: Profile Books.

PART ONE

CARING FOR THINGS THAT CARE FOR US

Care is part of our daily lives, but noticing and defining it can be elusive. Caring is a set of actions but is not merely a set of actions—nor is it merely an emotion. In human relationships, caring for another is a kind of emotional and practical labor that takes place within the context of a relationship of interdependence.

We delegate care work to objects when we create or enlist objects to support us and our projects, but this care work can become rote and cold if we don't find a way to encounter caring objects in an emotional register. Sometimes objects are designed to engage us in an emotional register, and at other times users create emotional ways of engaging with objects not designed with such an affordance. Often, to create things that can do care work for us in a way that feels caring to us, we must care for them as well.

1 PRIVACY AS CARE IN THE INTERNET OF THINGS

D.E. Wittkower

When a product contains an interface or a set of controls that do not function according to most users' mental models—for example, a numbered dial on a button meant to be pushed rather than turned—ideals of good design call upon us to fix the interface so that the product functions in the way the user expects. When a product presents itself as affording potentially harmful unintended actions—for example, an oven door that looks like it could be used as a stepstool to reach a high shelf—design considerations and engineering ethics call upon us to alter the interface so that these unintended actions are no longer afforded or to mitigate harm to the user that might follow from these unintended uses. Yet, when we speak of privacy issues, we tend to moralize about users rather than designers. Why is it that privacy issues are treated differently than other user intentions?

The overwhelming majority of contemporary discussion of privacy focuses on personally identifiable information (PII) as property, and research on privacy nearly always finds, in what is called "the privacy paradox" (Barnes 2006 and Norberg et al. 2007 are prominent early examples), that users claim to care about privacy but don't act in a way that effectively preserves their privacy. Barth and de Jong (2017) conducted an excellent meta-analysis of thirty-two "privacy paradox" studies, classifying ways in which paradoxical-seeming user behavior is accounted for by, for example, rational choice theory, public value theory, and symbolic interactionism. Kokolakis (2017) also conducted a meta-analysis, which surveyed twenty-two "privacy paradox" articles that used interpretive schemes ranging from behavioral economics to quantum theory. Neither meta-analysis found any paper that attributed the "privacy paradox" to problems on the design side rather than the user side, and neither meta-analysis found this omission noteworthy.

Research questions that inquire of users why they fail to act in accord with their beliefs implicitly validate industry design choices and terms of service, since they are treated as a given. The implication is that user behavior is the site where reforms are needed. If this issue were, however, treated like a normal case in engineering ethics, we would ask instead why design choices and terms of service fail to afford users means to act effectively with regard to their privacy concerns.

This fundamental bias disfigures much of the current debate about privacy in digital environments, and "privacy issues" as they appear in public, policy, and scholarly debate mostly participate in a process of responsibilization in which businesses' coercive extraction of user data is naturalized and excused, making protecting privacy a "user issue" in much the same way that sexual assault has been made a "women's issue." An effective and appropriate approach to social reform regarding sexual assault requires that we stop asking of those who have been assaulted what they were wearing or how they were acting and advising that they take better preventative measures, and that we begin asking instead about how the actions of those assaulting others have been facilitated and enabled. Similarly, an effective and appropriate approach to reforms around user privacy needs to stop judging users for "risky behavior" and shaming users for "oversharing" but should ask instead who is causing privacy-related harms and how those harms are being facilitated and enabled and seek to reform those structures.

This reconsideration of privacy issues should consider problems taking place on at least three layers: (1) poor design, (2) barriers to effective choice, and (3) divergent conceptions of the nature and value of privacy.

First, and most concrete, it may be that users fail to take effective action regarding their privacy concerns because of poor interface design or poor version control, or related design failures, where users may reasonably but falsely believe they have taken effective action to preserve privacy. It may be that instructions or settings are unclear or misleading, or that terms of service (TOS) change without notice, or that new features (e.g., Facebook's timeline) are introduced which violate established privacy-related user expectations. James Grimmelmann has written very well on these problems, describing Facebook as a "privacy virus" (2010), and advocating for using product liability law to make sense of privacy harms done to users (2009). Siva Vaidhyanathan (2012, 82–114) has addressed the related topic of designing interfaces to nudge users away from privacy with regard to Google, but in a way that has clear, broader application. Luke Stark (2016) has called for emotional engagement with users by designing for "visceral privacy" (2016, 23).

A broader problem of design is that industry's interest in privacy seems to be largely limited to gaining *pro forma* informed consent sufficient only to protect companies from liability or bad press, rather than actually designing to provide users with privacy in intuitive and expected uses of software or platforms. This concern is heightened by consideration of "dark patterns" of interface design, which not only fail to provide intuitive interfaces that support users' agency in acting to control their information flows but even nudge users away from privacy, creating an environment of digital hostile design. If the purpose of Facebook is, as Facebook says, to "give people the power to build community and bring the world closer together," then Facebook should be designed so that the user is not made responsible for assessing and mitigating risk prior to using the platform to realize this explicitly stated and intended function. For example, a bereaved user who is focused on the supposedly primary function of the platform, as he reaches out to friends for emotional support and shares memories of his childhood as he returns home to bury his father, should not be expected to step back and ask what kind of risk he's exposing himself to by revealing that he's traveling and therefore leaving his house unoccupied. The point is not that there is a simple or unproblematic design solution to allow for people to gain emotional support from one another without creating risk; the point is that Facebook's stated desiderata do not seem to be the goals toward which the platform's design aims and supports. Users' privacy is not designed for effectively and, in many cases of digital hostile architecture, is actively designed against.

Second, users may have no effective choice or no relevant alternatives allowing them to act in accord with their privacy concerns. Barth & de Jong (2017), very much to their credit, come very close to recognizing this structural problem in their discussion of Shklovski et al. (2014), saying that users accept TOS that undermine their privacy concerns because they "are resigned to the fact that they possess little power to change the situation" (Barth and de Jong 2017, 1049). It is only a small move from here to identify the problem not as "learned helplessness" (Shklovski et al. 2014) but as an actual power imbalance that exists outside of the user's mind, in which TOS are determined unilaterally and without avenues for negotiation, and in which there is no relevant alternative app or service which offers better or more transparent privacy provisions. This small move is, however, crucial, but it is underrepresented in these debates, as Barth and Jong accurately reflect. It is as if researchers keep asking why poor people who wish to be wealthy keep deciding to work low-wage jobs, without noticing that additional considerations may be relevant.

Third, it may be that the way users experience privacy does not map in scope, form, or texture onto the set of issues conceptualized as relevant to privacy by the current debate. Part of the problem is a fundamental difference between the way privacy is experienced in interpersonal relationships and the way privacy is recognized in legal and economic terms. As I've written previously (2016):

> A property-based understanding of personal information leads us to falsely think that "nobody cares about privacy anymore," and leads us to privacy education and advocacy efforts which are unlikely to succeed because they are mismatched with users' lived experience. It is unlikely to gain traction to emphasize how personal data is worth money and should be kept from circulation when the SNS [social networking site] user's lived experience is that personal information is what you tell your best friends when you need support, or is what creates intimacy and care—in other words, that personal information has (affective) value precisely through its circulation in personal and mutually beneficial relationships. (n.p.)

This chapter will continue my previous work on this topic. Previously, I've (2014) conducted a phenomenology of interpersonal information exchanges online in order to explore how publicity/privacy disperses into networks of information flows in multiply anchored selves, and I've (2016) used an analysis of inappropriate interpersonal behavior (being a lurker or a creeper) to provide a model for what users are likely to experience as inappropriate data access and use. Here I will conduct a phenomenology of privacy in interpersonal relationships in general and explore how those dynamics of privacy are likely to resonate in interactions within the Internet of Things. This will provide a set of privacy expectations native to users' lived experiences and an initial and regional exploration of where these do and do not map on to "privacy" as conceived of in the dominant legal-juridical privacy discourses.

A Phenomenology of Interpersonal Privacy

"Privacy" as it is recognized within primary Euro-American legal and moral conceptualizations is in a dichotomy with "public" and attaches primarily to concepts of rights (especially those held against governments) and economic interests (especially those in conflict with public goods) (Habermas [1962] 1989). This notion of privacy formalizes the positionality of a property-holding head-of-household and is thus a reflection of experiences of privacy historically pertaining primarily to straight, white, bourgeois men. Dominant discourses of privacy are based in patriarchal structures and male heterosexual experiences within these structures.

Women, children, and minorities have historically been disallowed from holding the kind of property to which dominant discourses of privacy pertain (e.g., businesses and sources of property, homes) and the moral rights of autonomy under political liberalism have only slowly and haltingly been extended to "private choices" which do not fit within the procreative heteronormative household. One example of the lingering and pernicious effects of the historically explicitly discriminatory basis of dominant Euro-American privacy discourses can be seen in the landmark case *Griswold v. Connecticut*, 381 U.S. 479 (1965). *Griswold* found that the right to privacy prohibited the state from a ban on contraceptive use but came to this finding on the basis of marital privacy rather than bodily autonomy. Providing autonomy from state interference on the basis of marital privacy provided rights on a discriminatory, heteronormative basis and complicated efforts to pass laws protecting married women from domestic abuse (Schneider 1990).

Innumerable other examples can be provided of how the legal-juridical model of privacy, dominant in Euro-American societies, has resulted in uneven and unequal protection as a legacy of the discriminatory history of property rights. We might consider how loitering, sit/lie laws, and laws against sleeping in cars have criminalized homelessness (Rosenberger 2017). We can consider how heterosexist laws pushed gay life out of protected spaces of home and business and into public spaces where queer activities and self-presentations can be criminalized, providing sexual privacy on a heteronormative basis (Chauncey 2014). We can consider how those receiving public assistance are subjected to intrusion and surveillance (Gilman 2008; Evans 2017; Smith 2018; Scalia 2019), since privacy is tied to private property rather than human dignity, resulting in the perception that impoverished and disabled persons should not expect, or even do not deserve, privacy.

The very specific experiences of the straight, white, abled male pursuing his own autonomy against governmental, economic, and cultural heteronomy are well-represented by the dominant rights-based legal-juridical privacy discourses. I seek to recover less valorized and less represented experiences of privacy in order to fill in missing pieces of the human experience of privacy. There is great value in exploring specific subaltern positionalities and how privacy appears from those standpoints, but the dominant privacy discourse is so thoroughly determined by property-rights-based concerns of heads-of-household that it will be sufficient for our purpose here, and will give us access to the broadest interpersonal experiences of privacy, simply to consider privacy in day-to-day interpersonal settings which are familiar to us all, regardless of our particular positionality and place within (or without) interpersonal relationships.

We will consider privacy as it appears in the interpersonal settings of parenting, friendship, romantic and sexual relationships, and in care for elderly and disabled persons. We will explore three themes across these relationships: (1) privacy and autonomy, (2) privacy and intimacy, and (3) privacy and consent. While the methodology here is phenomenological, this analysis is guided by insights from feminist ethics of care (e.g., Gilligan 1982; Noddings 1984; Slote 2007) and is especially indebted to Held (2006).

(1) Privacy and autonomy

Privacy desires and issues change in the parent-child relationship as the child grows older and becomes, first, more capable of self-maintenance, and, later, more capable of autonomy and self-determination of deeper sorts having to do with values and identity. Privacy is tied in this way to formation and performance of authority and control, where proper care from the parent requires a shifting negotiation of boundaries along with capabilities.

"Private" in the growing child's concern is not in a dichotomous relationship with "public" but serves as a delimitation of parental authority—the boundaries of parental authority negotiated within the family. The child's private business and the parent's authority over the child both fall within the private sphere as it is framed in opposition to the public sphere. It may be that "private" for the child means something different than "private" in the public/private dichotomy, or it may be that it means something quite the same, but within a different frame of reference or at a different level of abstraction (Floridi 2013, 29–52). In any case, the clear common element is that "private" designates a domain of autonomy—

autonomous choice as well as authority to control information flows across the boundary of what is designated as private.

"Paternalism" is the term usually used to refer to the purposeful reduction of the rights or freedoms of a presumptively autonomous individual for their own good (as perceived by those in a position to enact these reductions). In this context, in which we are recapturing the family dynamics which "paternalism" makes use of within a governmental level of abstraction (Floridi 2013), the gendered aspect of the term cannot be ignored. Should we speak, then, of "maternalism" as well as "paternalism" within the family? We can certainly see different and gendered discourses of control and autonomy within the family. In the register of justice, we might think of claims of the child that they are owed greater autonomy, such as "it just isn't fair" or "you let sibling [X] do [Y] at [Z years of age]." Parental rebuttals having to do with justice and what is owed to others are significantly but not exclusively male-gendered: "not as long as you're living under my roof" or "because I said so" both point to the authority of the head of household and tie justification to that typically and historically male heterosexual positionality. We can also easily call to mind children's claims based in care rather than justice, such as "you have to let me make my own choices" or "I need to be able to figure out who I am on my own," as well as replies that similarly reflect care rather than justice and property rights: "I'm just worried about you," or "you'll always be my baby."

In either case, and whether or not these gendered ways of negotiating the shifting boundary of the child's privacy and relative autonomy are well-named "paternalism" and "maternalism," what we observe is a dynamic where privacy emerges de novo as a child's realm of autonomy within and extending beyond the parent's realm of autonomy, whose boundaries are negotiated by concerns of both justice and care, and the proper extent of which is tied to the ability to make responsible, informed choice. Trust and prudence must be used in bringing this boundary to accord with the child's abilities—trust in allowing the child autonomy even when it extends beyond the guardian's comfort or concerns choices where there is disagreement, and prudence when the child feels entitled to autonomy for which they are not prepared. The child's abilities to be autonomous, furthermore, are ceteris paribus to be maximized by guardians—it would be both uncaring and unjust to fail to encourage a child to develop the independence of mind and the breadth of knowledge that allows for a child to deserve privacy, as for example in the systematic infantilization of women that has gone part and parcel with women's political and social disenfranchisement in Euro-American societies.

Later in life, the child must come to terms with taking authority over the parent as her parent's world becomes smaller and their autonomy diminishes, and her parent's privacy must be sacrificed in order to provide care (Fig. 1.1). As with the child, the ability of the elder to be self-determining is ceteris paribus to be maximized, and purposefully diminishing autonomy seems clearly unjust and uncaring—as for example in the too-common case of off-label administration of antipsychotics to dementia patients in order to make them easier to manage in elder-care facility settings.

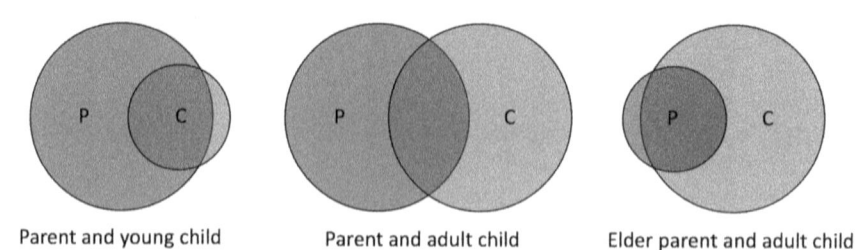

| Parent and young child | Parent and adult child | Elder parent and adult child |

FIGURE 1.1 Realms of privacy as overlapping ranges of effective executive autonomy within an interdependent relationship.

It is important to make explicit, though, that maximizing *ability* to be self-determining seems to be a moral imperative in both child rearing and elder care, but maximizing *actual* self-determination certainly is not. Forcing a child to make it on her own as soon as she is able is uncaring, and less than a child deserves morally if not legally; and placing a parent in an elder-care facility may provide more autonomy than personally caring for that parent, but it is not necessarily a just or caring choice. Furthermore, even at these moments when we often feel much is at stake in gaining or maintaining autonomy, we value our interdependence with those we care about and care for.

We shouldn't seek to maximize privacy/autonomy, but to maximize the *ability* to be autonomous and maintain privacy. This allows for the greatest proportion possible of diminished privacy and increased dependence to be voluntary, actively chosen, and desirable.

Here we see another strong difference from how "privacy" appears in dominant legal-juridical discourses, where privacy is to be hoarded: in interpersonal relationships, we regularly voluntarily "give away" our privacy and take pleasure in doing so. We talk about this dynamic in terms of "intimacy," which describes slightly different things as a state and as a process. This will be the second element of this phenomenology of privacy in interpersonal contexts.

(2) Privacy and intimacy

I've written previously (2016) about how the value of private information to interpersonal relationships has to do with its circulation as well as its withholding. Certainly there are things we prefer others not to know, but in the affective economy of interpersonal information flows, scarcity drives up value, and we generate bonds by demarcating levels of progressively more restrictive access. While I'm not prepared to say that this is the full or only meaning of the term, or that this captures its essence, this method of creating affective charge through voluntary and selective diminution of privacy is what I will here call *intimacy*.

Intimacy is a felt experience, but it inheres in the relationship rather than in either or both parties, and it is not mechanistically produced through provision of otherwise private information. To tell someone something private outside of a relationship of mutual and progressively more private information exchange is not to create intimacy, but instead is likely to feel inappropriate, or a kind of imposition, since it makes an unbidden demand of trust from the one confided in. When "intimacy" is felt as a unilateral emotion, when intimacy is not present within the relationship, this experience of pseudo-intimacy is better called "fixation" or "fantasy," and if disclosed, it is not likely to create intimacy but instead a creepy feeling (Wittkower 2016). So, even though intimacy is a felt experience, it is based on a history of trusted bilateral information exchanges, and to that extent is well- or perhaps even better-described in terms of affective economy.

We can have intimate moments in all sorts of relationships, and close relationships of all sorts include intimacy to greater or lesser degrees, but we nonetheless name something distinct when we refer to an "intimate relationship"— typically a bilateral dyadic flow of information of a sexual nature leading to "being intimate" as a euphemism for sex. Intimate relationships, however, are not distinct only because of the sexual nature of information exchange, and it is not incoherent to consider a romantic relationship in which no sexual activity ever occurs to be an "intimate relationship." This, of course, leans heavily on "romantic" to define necessary and sufficient conditions for an intimate relationship, but thankfully we need here only to outline and characterize these experiences rather than fully define them.

In functional terms, intimate relationships are intimate in that the sharply delimited exchange of information that creates intimacy is *constitutive* of the relationship. When intimacy-creating information received within an intimate relationship is disclosed to third parties, the relationship is threatened or broken. When one partner of an intimate relationship enters into another intimate relationship, the first relationship is typically threatened or broken. When the intimacy of an intimate relationship declines through distance or emotional withdrawal, the relationship is threatened or broken.

Intimacy, further, is subject to constant negotiation and renegotiation. In any kind of relationship, intimate or not, the exchange of private information does not constitute intimacy unless it takes place in the context of bilateral consent— and even in an intimate relationship, consent for ongoing exchange of private information is never settled but always requires affirmation. This will be the third and final element of this phenomenology of privacy in interpersonal contexts.

(3) Privacy and consent

Privacy creates, requires, and is required for autonomy, but we willingly give up privacy (and autonomy) in order to create interdependent and intimate bonds with others—most markedly within intimate relationships. The consensual nature of this bilateral diminution of privacy is central to the creation of intimacy and mutual interdependence. The same information exchanges which give weight and meaning to our relationships—even fleeting intimate moments between friends—would be violations of trust, creating threat and coercion, when outside of the context of mutual and caring relationships of established trust.

Negotiation of privacy as intimacy also provides a model of consent strikingly different from the model used in legal-juridical contexts—in interpersonal relationships, even outside of romantic relationships, consent is always temporally and contextually specific, requiring different degrees of refreshed consent depending on the level of intimacy involved. In sexual activity, at one extreme of intimacy, consent must be constantly refreshed through verbal or nonverbal cues, and in activity requiring high levels of trust, specific systems to assure consent are used, such as "safe words" in role play or BDSM sexual activity. In less intimate relationships, such as friendship, consent must still be negotiated on an ongoing basis; a truck once used to move a sofa should not be assumed to be available whenever the need arises in the future, and having someone over for dinner certainly doesn't imply an ongoing open door policy as night falls each evening.

This everyday ongoing negotiation of boundaries in and through which we maintain relationships of care with care also requires us to inquire beyond express consent. If we are fortunate in our relationships, we have some friends and family that will help us even at great personal cost, and we should not trust their answer when we ask, "Are you sure it's no bother?," nor should we act as if the favor done is as trivial as they pretend it to be. Imposition on someone exhibiting such giving behavior surely becomes abusive at some point, even if it remains nominally consensual. When we consider our own choices, though, it may be best sometimes to act as if, for example, helping a friend or a parent were purely consensual, even if we feel sometimes a bit coerced or constrained in our choice. Interdependence requires flexibility and compromise, and personal relationships are not always amenable to objective debate in order to agree upon fully voluntary and mutually beneficial agreements.

Application to Five Cases in the Internet of Things

In the above phenomenology of privacy in interpersonal relationships, we saw (1) how privacy was necessary for and to some extent constitutive of autonomy, (2) how we create intimacy and interdependence through controlled breaches of privacy, and (3) how these interpersonal relationship-building functions of information flows require constant negotiation of consent. Now we can ask how this revitalized understanding of the human experience of privacy helps to reevaluate how people experience privacy concerns.

Each of these aspects of the interpersonal experience of privacy provides useful guidance in thinking about how things and systems can relate to us. Here we will apply them to systems in the Internet of Things (IoT), especially domestic technologies and those which play roles of support in our everyday lives. These technologies are ideal for exploring a care-theoretic model of privacy due to the care-like work that they perform; the physical presence that they have in familial, household spaces; and the personal and personified interactions we often have with them. A variety of kinds of IoT devices and systems will be discussed, chosen to represent a diversity of kinds of interactions and relationships of these kinds within the Internet of Things: GPS navigators, the Amazon Alexa virtual assistant, Nest, and two medical robots—PARO and RIBA.

A. GPS

Since we frequently turn them on in the driveway before departing the home or use them to get directed to our home (and don't turn them off while en route), GPS systems rigidly insist in telling us how to drive in and out of the neighborhoods we live in, every time, year after year. By failing to withdraw (Ihde 1990, 109) when their help is unneeded, they invade our privacy after the fashion of a parent who refuses to leave one's child space to make their own choices. The GPS system does not remove our autonomy, but it inadvertently belittles us and disrespects our knowledge and judgment by treating us as if we know nothing about our own home.

This finding from the above phenomenology identifies a privacy experience that doesn't show up within current privacy discourses, but I doubt that this finding is of any great importance. This demonstrates strengths and weaknesses of the approach taken in this chapter—tracking privacy as we experience it in human interaction can get us to experiences and concerns that a legal-juridical approach will not recognize, but those concerns may not line up well with technical or legal problems and thus may not be easily addressed or resolved, and may not have significant enough consequences to be worth resolving.

B. Digital Assistants

Virtual digital assistants, like Amazon's Alexa, are interesting in the context of peer-privacy concerns, such as those that emerge in friendship or romantic relationships. Our relationship with Alexa is altered as skills are discovered and installed—intimacy is created as the user gets to know more about Alexa and installs skills, creating interests and concerns in common. This creates an experience akin to a relationship deepening through shared activities and mutual recognition. This experience is, however, belied by most virtual assistants' inability to distinguish between users, or learn names, and by their default market orientation and proclivity to offer to buy stuff on your behalf. Further, the motivation of opening up to persons in order to create bonds and affective valences can play little role with virtual assistants, since their "personality" consists of little more than Easter eggs and silly replies to requests—for example, that Alexa sing a song.

Despite these inhibiting factors, virtual assistants can't help but be experienced with an aura of intimacy, simply because of where they appear physically and socially. Physically, they are often in kitchens and other spatial foci (Borgmann 2009) of the home and present in backstage (Goffman 1959) environments and situations, bringing us the morning news as we prepare our coffee or pack the kids' lunches. Socially, they become interactants and objects of play for children and information gateways for adults. Crucially, the humans which these assistants are assisting must develop a theory of mind of the virtual assistant in order to use them properly (Wittkower forthcoming), and a controlled vocabulary that describes the mental objects and processes in this theory of mind—for example, in order to get the right result, I have learned to specifically say, "Alexa, ask NPR One to play the latest hourly newscast," and my daughter, for her part, has learned to append "original motion picture soundtrack" to her requests that Alexa play *Moana* or *Equestria Girls*, despite the fact that this is not an intuitive way for kindergarteners to refer to movie music.

C. Home Automation

Home automation systems like Nest similarly customize to the user and take central positions in intimate physical spaces but do so in a way that withdraws from experience (Ihde 1990, 109) rather than through shared experiences. Like a parent, they demonstrate care, stepping in to help in a way that expands rather than reduces the autonomy of their charge—but this caring relationship is strongly in conflict with analysis and resale of aggregated user data, possibly producing a stronger moral complaint than similar systems which take place on a more clearly economic playing field, like customer loyalty cards that allow users to save on groceries in exchange for data on buying habits.

Heather Wiltse, in her role as editor of this volume, also points out that "'smart home' devices can also be used to mediate domestic abuse." As reported in the *New York Times*, numerous cases have been documented where abusive men have maintained or regained technical control of home automation systems, controlling lights, thermostats, and music within former residences occupied by estranged partners. It may also be possible to stalk former partners through distant access of home security video systems (Bowles 2018). Wiltse insightfully comments, "This is another scenario where the heteronormative and patriarchal foundations of privacy discussions are on full display, and so even here being concerned only about commercial use of collected data misses the arguably more fundamental need for integrity, safety, and control in one's home environment."

D. PARO

PARO, a robot pet-therapy seal, develops a custom, learned "love language" of affectionate acts that it prefers to receive from its human companion, developing an idiosyncratic interactive and intimate relationship with its human companion.

PARO also demonstrates need and helplessness to its human companion, allowing elderly patients with dementia to be needed and to take on caretaker roles, evoking a feeling of autonomy and independence by contrast—experiences which may not be empowering, since they don't reflect actual gains in autonomy, but which are positive and stabilizing.

Experiences and feelings can't be expected to fit neatly into conceptual categories, nor should they be expected to emerge rationally. Here, we see the interdependent experiences of independence, autonomy, and privacy bound up with the aged adult's fading roles of authority and caregiver. I wonder whether these entangled concerns and their attendant feelings are what is often referred to as "dignity" when discussing patients, especially patients who are becoming increasingly dependent upon persons and technologies. If so, then we might say that PARO provides a benefit to some patients because it allows them to experience giving care and having another depend upon them, providing variation and leavening to their increasing dependence in other aspects of life—a mixture of depending on others and being depended upon that characterizes normal life, rather than the exceptional circumstances of what may at least seem to be pure patiency associated with infancy and with the end of life.

E. RIBA

Patients who are unable to be self-sufficient and autonomous physically rather than psychologically may be served by RIBA, a large teddy-bear-headed robot designed to gently lift, carry, and set down humans. By sourcing mobility assistance to RIBA rather than a human, patients can maintain privacy while receiving needed support to, for example, get out of bed, or get on and off the toilet. In this case, where the user enters the relationship from a position of dependence and disability, if the user is able to gain support from a relationship which is less humanized, robust, and interactive, this may be preferable to users (Borenstein and Pearson 2010, 282), since dependence is tied to a loss of privacy, autonomy, independence, and self-determination which is very uncomfortable or unsettling when not within the context of an ongoing personal relationship of care and trust, and often enough undesirable even within such a relationship. RIBA represents not only decisional autonomy but a close proxy for executional autonomy as well (Fine and Glendinning 2005, 610) through a sort of hybrid embodiment/alterity relationship (Ihde 1990) enabled by the ways that RIBA is not a "smart" device. A RIBA that was too interactive or that "learned" much about its user would take on too many social elements and would tend to become destructive of rather than preserving of user privacy.

Conclusions

We can draw several conclusions from this reconsideration of privacy in IoT, using an interpersonal understanding of privacy rather than a legal-juridical one.

First, we note that different issues come to the fore. Privacy concerns in a legal-juridical conception of privacy view all diminution of privacy as a loss: a necessary evil at best. In interpersonal relationships, privacy is something we enjoy along with others, and we enjoy and seek out "losses" of privacy in the form of intimacy and care. Interpersonal privacy doesn't appear as a value inhering in data that is maximized by withholding and protecting, but instead privacy appears as a negotiated boundary across which personal information flows, producing affect and emotional value through its bounded circulation. Perhaps most unexpectedly, we see how closely tied privacy and self-determination are in interpersonal contexts, and the benefit of privacy and autonomy in some contexts, but the value of interdependence and mutual determination in other, more intimate relationships.

Lived interpersonal experiences of privacy signal other things that users may prefer or even intuitively expect in technical contexts, such as the revisability of consent. Consent in a legal-juridical context is a box to be checked, resulting in the waiving of rights—neither of which are at all similar to consent in an interpersonal context. Managing and meeting user expectations require that data use policies reflect users' mental models, and when these more interpersonally based user expectations are reinforced by technologies like virtual assistants that have a social presence and that seek to act as partners with the user, it is all the more important that they approach user data in a way that fits with interpersonal interpretations and mental models of consent and data management, perhaps even to the extent that they should act as "privacy allies," as suggested by Diane Michelfelder in her chapter in this volume. Businesses can't have it both ways: if they want us to approach their tech socially, they need to appropriately socialize their tech.

Finally, the connection between privacy and intimacy provides useful direction as well. Trading privacy for increased customization of services is a natural step when it appears in a social context of "getting to know you and what you like," but this only feels comfortable if markers of intimacy are included. Central to that experience of intimacy are trust, care, and faithfulness. Any company producing a good or service which is designed to create an experience of intimacy should keep faith with the user experience that their product's success depends upon by creating objects and platforms that exhibit care toward their humans.

References

Barnes, Susan B. 2006. "A Privacy Paradox: Social Networking in the United States." *First Monday* 11 (9): n.p.

Barth, Susanne and Menno D.T. de Jong. 2017. "The Privacy Paradox—Investigating Discrepancies between Expressed Privacy Concerns and Actual Online Behavior—A Systematic Literature Review." *Telematics and Informatics* 34 (7): 1038–1058.

Borenstein, Jason and Yvette Pearson. 2010. "Robot Caregivers: Harbingers of Expanded Freedom for All?" *Ethics and Information Technology* 12 (3): 277–288.

Borgmann, Albert. 2009. *Technology and the Character of Contemporary Life: A Philosophical Inquiry*. Chicago: University of Chicago Press.

Bowles, Nellie. 2018. "Thermostats, Locks and Lights: Digital Tools of Domestic Abuse." *New York Times*, June 23, 2018. https://www.nytimes.com/2018/06/23/technology/smart-home-devices-domestic-abuse.html.

Chauncey, George. 2014. "Privacy Could Only Be Had in Public." In *The People, Place, and Space Reader*, edited by Jen Jack Gieseking, William Mangold, Cindi Katz, Setha Low, and Susan Saegert. New York: Routledge, pp. 202–206.

Evans, Dominick (@dominickevans) et al. 2017. "EVV (electronic visit verification) goes into effect on Jan 8 …" *Twitter*, December 28, 2017. https://twitter.com/dominickevans/status/946495094437433344

Fine, Michael and Caroline Glendinning. 2005. "Dependence, Independence or Inter-Dependence? Revisiting the Concepts of 'Care' and 'Dependency.'" *Ageing & Society* 25 (4): 601–621.

Floridi, Luciano. 2013. *The Ethics of Information*. Cambridge: Oxford University Press.

Gilligan, Carol. 1982. *In a Different Voice: Psychological Theory and Women's Development*. Cambridge, MA: Harvard University Press.

Gilman, Michele E. 2008. "Welfare, Privacy, and Feminism." *University of Baltimore Law Forum* 39 (1): 1–25.

Goffman, Erving. 1959. *The Presentation of Self in Everyday Life*. Garden City, NY: Doubleday.

Grimmelmann, James. 2009. "Privacy as Product Safety." *Widener Law Journal* 19: 793–827.

Grimmelmann, James. 2010. "The Privacy Virus." In *Facebook and Philosophy: What's on Your Mind?*, edited by D.E. Wittkower. Chicago: Open Court.

Habermas, Jürgen. 1989 [1962]. *The Structural Transformation of the Public Sphere*. Translated by Thomas Burger. Cambridge, MA: MIT Press.

Held, Virginia. 2006. *The Ethics of Care: Personal, Political, and Global*. Cambridge: Oxford University Press.

Ihde, Don. 1990. *Technology and the Lifeworld: From Garden to Earth*. Bloomington: Indiana University Press.

Kokolakis, Spyros. 2017. "Privacy Attitudes and Privacy Behaviour: A Review of Current Research on the Privacy Paradox Phenomenon." *Computers & Security* 64: 122–134.

Noddings, Nel. 1984. *Caring: A Feminine Approach to Ethics and Moral Education*. Berkeley: University of California Press.

Norberg, Patricia, Daniel Horne, and David Horne. 2007. "The Privacy Paradox: Personal Information Disclosure Intentions versus Behaviors." *Journal of Consumer Affairs* 41 (1): 100–126.

Rosenberger, Robert. 2017. *Callous Objects: Designs Against the Homeless*. Minneapolis, MN: University of Minnesota Press.

Scalia, Kendra. 2019. "Electronic Visit Verification (EVV) Is Here." *Disability Visibility Project*, March 24, 2019. https://disabilityvisibilityproject.com/2019/03/24/electronic-visit-verification-evv-is-here/

Schneider, Elizabeth. 1990. "The Violence of Privacy." *Connecticut Law Review* 23 (4): 973–1000.

Shklovski, Irina, Scott Mainwaring, Halla Hrund Skúladóttir, and Höskuldur Borgthorsson. 2014. "Leakiness and Creepiness in App Space: Perceptions of Privacy and Mobile App Use." *Proceedings of the 32nd Annual ACM Conference on Human Factors in Computing Systems*, 2347–2356. New York: ACM.

Slote, Michael. 2007. *The Ethics of Care and Empathy*. New York: Routledge.

Smith, S.E. 2018. "Electronic Visit Verification: A Threat to Independence for Disabled People." *Rooted in Rights*, July 31, 2018. https://rootedinrights.org/electronic-visit-verification-a-threat-to-independence-for-disabled-people/

Stark, Luke. 2016. "The Emotional Context of Information Privacy." *The Information Society* 32 (1): 14–27.

Vaidhyanathan, Siva. 2012. *The Googlization of Everything: (And Why We Should Worry)*. Berkeley: University of California Press.

Wittkower, D.E. 2014. "Facebook and Dramauthentic Identity: A Post-Goffmanian Model of Identity Performance on SNS." *First Monday* 19 (4): n.p.

Wittkower, D.E. 2016. "Lurkers, Creepers, and Virtuous Interactivity: From Property Rights to Consent and Care as a Conceptual Basis for Privacy Concerns and Information Ethics." *First Monday* 21 (10): n.p.

Wittkower, D.E. Forthcoming. "What Is It Like to Be a Bot?" In *Oxford Handbook of Philosophy of Technology*, edited by Shannon Vallor. Cambridge: Oxford University Press.

2 ATTACHMENT TO THINGS, ARTIFACTS, DEVICES, COMMODITIES: AN INCONVENIENT ETHICS OF THE ORDINARY

Michel Puech

Where Is the Problem, If Any?

We live in a world of things, artifacts, devices, and commodities, and we take them for granted, in a sense: their presence and their functions remain in an existential background which is typically transparent, as long as everything runs smoothly. However, when questions arise about buying a new mobile phone because I simply have a crush on it, while the "old" one bought one year before is still perfectly working, or similar cases, this background comes to the front end of conscious life and requires a specific valuation and decision frame of reference. My first point is to humbly assume that this frame of reference is often lacking, blurred, or inconsistent. In these situations the professional philosopher might experiment that our "applied ethics" does not apply to real ordinary things beyond the "extraordinary" (trolley) case studies in our textbooks. When philosophers embrace these kinds of ordinary problems, they reach the conclusion that the nearest (physically, functionally, emotionally) *things* are elusive. Ordinary things are ontologically vague and morally thin in the best case; they remain unseen and tacitly despised most of the time. When their valuation or a decision about them is needed, we seem to follow some sort of hazy "intuition" deprived of explicit argument. The very personal management of ordinary objects tends to remain vague and obscure because we know how poor our rationale would be if we had to formulate and defend it. It happens by nature in intimacy, in Wittkower's sense (see his chapter in this book), because ordinary things belong there, by nature, I think, or by design, according to social and human sciences research.

There is a brilliant analysis of our engagement in "ordinarity" all along Heidegger's *Being and Time* (Heidegger 1976 [1927]). It goes in terms of *Durchschnittlichkeit*, Averageness: the "ordinary" consideration of things, which remains instrumental and "ontic," as opposed to the "ontological" approach, which leads Heidegger to grandiose views on time and death. Being deeply immersed in the ordinary of life is the "natural" attitude, as opposed to the dignity of the aristocratic detached point of view suitable for philosophy. Everydayness then is only mentioned as a derelict mode of human existence, which must be abandoned to reach authenticity. A tradition of technophobic philosophers still sustains this view of everydayness as undignified, for existential reasons (dereliction) or political (capitalist conspiracy manipulating consumers) reasons, or both, as they are compatible in mainstream "social critique."

I believe on the contrary with Albert Borgmann (Borgmann 1984) that everydayness can be positively addressed and that a form of authenticity can be claimed for it, or better: can be reached within it. And I suggest an ethics and even a wisdom for giving shape to this mode of life, theoretically, and practicing it in personal pursuit of practical wisdom.

The Heideggerian discrediting attitude toward the possible involvement with, attachment to, or emotions for things has deep roots in Western culture—in Cicero, an author who is not much read today but whose common sense judgments found many of our consensual views. It is "absurd," says Cicero in his *De amicitia* (first century BCE), to "love" what he calls "inanimate things" for a very simple reason: they cannot "love you back" (Cicero 1923, chap. 14). Both premises can be denied, I dare say: on the one hand, there is no need of reciprocity ("loving back") in love, alas, and on the other hand, things *can* love us back, in their own way, which is not so despicable, as I hope to show. Some forms of this loving back are manipulative marketing tricks (Alexa or Paro, mentioned in this book, *tamagotchi* in this chapter) in their origin, or "ontologically," but they can be seen from a candid "ontic" point of view as real attachment partners, in a subtle, slightly ironic, but nevertheless very common behavior—I know that this cake/telephone/app has been carefully designed and tuned to seduce me but I love it anyway, keeping in mind that there may be a hidden agenda or script in it. Paraphrasing Kranzberg's law of technology: attachment to things is neither good nor bad; nor is it neutral. The conclusion is well known: then it requires an ethical assessment.

Ethical Considerability

In the recent history of applied ethics there is a visible trend to continually expand ethical consideration to new ontological domains, moving in concentric circles toward larger realms (Verbeek 2005; Turkle 2007, 2008; Bogost 2012). It can be summarized in the following series: considering as ethical subjects (bearer of value, deserving ethical attention), successively: every human and not just me, my tribe, race, gender; then animals, plants, and ecosystems. Can this expansion reach ordinary things, including digital objects or any mundane commodity? It must, if we, philosophers, are willing to contribute to the social conversation on the so-called digital addiction or the kind of "trust" that algorithms deserve to rule our lives (as Google Maps paves our way) and comparable topics. The social demand is increasing on a lot of issues that can be characterized as ethical questions bearing on the relationship to "things" because our moral systems have been conceived for "humans only" relationships.

Concerning ecosystems as a whole and natural resources, like water or sand, environmental philosophy successfully argues that nonliving things can be ethically considerable. Now an interesting line of extension, beyond the *human/living/nonliving* first stages of transition, is the step from nonliving things to nonmaterial "things" (see Heather Wiltse's chapter in this book) and to our digital "assistants," personal assistants, or home assistants, who dwell in the infosphere and are talking to us through *access things* like the smartphone or a GAFA (Google Apple Facebook Amazon) home device like "Google Home" and Amazon's "Echo"—on the latter see Diane Michelfelder in this book and the heartfelt appeal in the *New York Times* (https://nyti.ms/2vauqDU, November 7, 2017) "Alexa, where have you been all my life?," stating: "How a sleek, smooth-talking cylinder from Amazon stole our hearts, bamboozled our spouses and enchanted our children."

In response to this challenge, can we reiterate something like the arguments for the ethical considerability of nonwhites, nonmales, nonhuman natural entities and implement them to every "thing" in human life, my focus being on the most ordinary and humble? Is there any need to do so? Obviously there is, because of the attachment bonds existing between humans and some of these things, bonds that we need to be aware of and that we need in the end to morally endorse or decide to censure.

These bonds are made of care, attachment, and perhaps love (Dumouchel and Damiano 2017). In our relation to objects, we take care of them, we care about them, and we care for them. Maintenance and repair already belong to an ethic of care suggests Steven J. Jackson: "Finally, foregrounding maintenance and repair as an aspect of technological work invites not only new functional but also moral relations to the world of technology" (Gillespie, Boczkowski, and Foot 2014, 231). Ordinary objects are part of what he calls the "technological." In his early and influential (in the technological milieu) book about Zen and motorcycle maintenance, Pirsig (1974) made it clear that relating to a "thing" like a motorcycle opens a vast sphere of strong, delicate, and unquestionably moral experiences, which graciously or grotesquely blend themselves into human existence, shape it, and transform it.

One more step deep into this kind of existential experience (for some it may be a leap of faith out of mainstream social critique) is required to come to the idea that some things can take care of us and that they care about us, in a certain sense, as much as we care about them. This issue was brought to the public attention in the 1990s with the *tamagotchi* phenomenon: egg-shaped "digital pets" invaded the world of toys and this new kind of interactive toys deserve the title of existential companion, like a living "pet." The *tamagotchi* needs the care of its master/owner/companion, and it even begs for constant attention. This toy is already a paradigmatic attention-catcher, preluding the smartphone area, and parallel to the Pokémon fever, which remains an outstanding case study of attachment to "things" that have a very vague ontology. Pokémons are fictions but their world is so coherent that it is reassuring for the children engaged into Pokémons chase and management. The augmented reality "Pokémon Go" fever in 2016 (see Galit Wellner's chapter in this book) combines different experiences of attachment: to the creatures themselves, to the smartphone "through which" they are accessible, and incidentally to the real-world places where to chase virtual Pokémons—these places (parks, historical city centers) being rediscovered as finally worth visiting by the youngest generation of millennials.

There were research projects about "affective computing" in the 1990s, some of them suggesting a "wearable affective agent" to implement further new generations of these "agents that learn your preferences" (Picard 1997, 101). Embarrassingly, when they have learned your preferences, the now existing intelligent algorithms of the infosphere can modify your preferences, or use them, or sell them. "Affective" in Picard's work means "intelligent" like in "artificial intelligence" and in the commercial commodities we are now considering the affective bond with humankind is manipulative by design, at least as an attention-catcher.

After that came the time of a rapidly expanding literature on robot and AI ethics, culminating with luridly titled articles like "Do You Want a Robot Lover? The Ethics of Caring Technologies" by Blay Whitby, in Lin, Abney, and Bekey (2014, chap. 15). When the question is posed at the level of sex, ethical considerability is taken for granted, because sex is a prominent inter-human ethical issue. Aiming at a large audience, there was an online test in April 2017 by the *New York Times*: "Are you in love with your phone?" (https://nyti.ms/2psvbYX). The question is actually not about the facts (our behavior relating to our mobile phone) but about the moral acceptability of the fact. It can be rendered as: "Do you endorse the fact that you are in love with your phone?" and perhaps: "Do you have another and more appropriate qualification for what you feel in the intimate relationship with your phone?" From a philosophical point of view, the expected negative answer is due to the lack of ethical concepts concerning our attachment to things and to technological things in particular. This is embarrassing when we try to inquire into caring for things and possibly discrediting if philosophers have nothing to say for or against "robot lovers," in the romantic sense of "love" I mean. If we have to live with Hollywood's views of a loving attachment to an AI or robot, we might never escape murkiness and prejudices.

Attachment Theory Relating to Things

I believe that "attachment theory" can help to conceptualize the ethics of our relation to objects. This theory was born in the mid-twentieth century in the field of interpretative psychology. As it is a precise and rather complex theory, an introductory Wikipedia perusal is not useless, but the robust scientific literature supporting the theory should be browsed (see, for instance, Cassidy and Shaver 2002). Attachment theory is now an evidence-based medical science and a soaring therapy practice. Naturally, my philosophical approach does not meet the requirements of this science and this practice, but my purpose is simply to point to potential innovative methods in the ethics of things. Ironically, attachment theory bears entirely on human-to-human relations, originally, but for no good or definitive reason, I will argue. Its focus is on the (human) child's relation with a (parental) caregiver, the "attachment figure." In this vocabulary my problem is to address "attachment objects" (nonhuman).

My argument is based on the fact that the "humans only" limitation of attachment theory is not well-founded. Let us rewind back to one of the most important founding papers in the theory, Harry F. Harlow's article "The Nature of Love" (Harlow 1958). Harlow's untenable prejudice against objects remained strangely unnoticed. He is testing the behavior of baby monkeys in comparative experiments with their real mothers and some "surrogate mothers" made of diverse materials. The most successful surrogate mother is "a block of wood, covered with sponge rubber, and sheathed in tan cotton terry cloth," which is to me indisputably a "thing," a nonliving material object. Attachment theory quotes Harlow's experiment again and again, but no one (as far as I know) ever stressed the *artifact* dimension in this animal model experiment. Conclusions bear on humans-only relationships; the fact that what is observed is a monkey-*thing* relationship seems not to matter. It matters, not on the "monkey" side (for me) but on the "thing" side of the relation. In phenomenological "variations" experimenting with human subjects, we can substitute as "surrogate attachment figures" quite a lot of objects: personal computer and smartphone, cars, software ("local" and online), home comfort commodities (for food, sleep, body hygiene, distraction, etc.).

Considering a human-thing relationship, the relevant inputs from attachment theory must remain the same as in interpersonal relations:

(1) Interacting with a *caregiver*, to regulate feelings, in particular to manage situations of "alarm," danger, discomfort, stress.
(2) The caregiver (1) provides protection and emotional support.
(3) Proximity to the caregiver is identified as a stable resource for existential support (2).
(4) This "secure base" (3) encourages exploration of the world and it allows self-construction in a "goal-corrected partnership" with attachment figures.

The whole pattern is easily applicable to ordinary things and artifacts in order to expose them as attachment-related objects.

(1) They provide "care." The home is the first and best caregiving resource, for the sick and elderly person as well as for every human being. Domestic appliances (fridge, couch, coffee machine) deserve the same status, but also Google Maps when one is lost, because it really and aptly cares, as any telephone will do in case of accident, sadness, or any existential alarm.

(2) They provide reassurance and emotional support. We better understand how they can since Winnicott's theory of transitional objects (Winnicott 1971), the comforter toy or blanket without which the child would never fall asleep and which has the magical power to bring solace and consolation for every small or big sorrow in a child's life. For adults, some familiar or special garment provides reassurance and emotional support, or it may be some familiar food when one is abroad, and also one's car (or motorcycle), or familiar radio or TV news channels (when far away from home you find the local radio of home on the Web and fall asleep with it: this is Winnicottian).

(3) Their proximity is a "secure base," made of ordinary things and artifacts which are attachment-related objects, such as the familiar environment of a home, the proximity of the smartphone in one's pocket, bag, at one's bedside: within arm's reach. Through familiar electronic devices, there is a virtual proximity of real friends or Facebook "friends." Proximity makes sense, then, when considering one's intimate infosphere.

(4) This secure base incites to explore: suitable clothes for diving into foreign social circles, running shoes inciting to run (attachment to one's sport gear), and of course the smartphone inciting to explore the universe, like the Star Trek multifunctional "tricoder," telling you instantly if the atmosphere on this planet is breathable, or more prosaically if there is any fast food in the area.

Understanding our relation to objects can draw from the diverse *attachment styles* in the theory. They are behavioral patterns for coping with life problems, particularly in social and personal (human) relations. In an oversimplified interdisciplinary version we just need to mention the main distinction:

(a) Secure attachment styles: the caregiver is reliable; there is a stable secure base; this secure base (its available proximity) incites to actively explore the world; and in human physiology the attachment system (an identified specific biological circuit) operates efficiently, and most of the time it remains unactivated, in the background.

(b) Insecure attachment styles: the caregiver is not competent enough, not available enough, or nonexisting at all; then exploration capacities, emotional and communicative capacities are impaired; and in the end the instability of reassurance resources leads to a range of anxiety, ambivalence, and "avoidant" attitudes, keeping the attachment system in a perpetual alarm.

What comes next is providential for an ethics of objects: the theory of *attachment disorders*. It is a non-simplistic theory, because insecure attachment is an existential style, not a disorder in itself; the theory and its derived therapies insist on this premise. But some existential disorders are attachment-related, particularly those pertaining to the attachment style called "insecure disorganized." This last category allows philosophers of technology to describe in new terms some poorly defined and poorly understood "pathological" relationships with objects.

FOMO, the "fear of missing out" (compulsive message checking), is connected to the typical *fear of loss* in insecure attachment, an original and basic concept in attachment theory. The most important book in the theory is *Attachment and loss* (Bowlby 1969).

Addiction to video games or to online porn can be interpreted as inappropriate looking for a reassurance that the *caregiving object* is not able to provide, then compensating the frustration with obsessional excess, plus withdrawal symptoms, and an impairment of one's openness to the world.

Benign and severe technophobia, resistance to technological change, qualify as symptoms of difficulties in investing new attachment objects and in extending and adapting one's secure base (of attachment objects), which lead to a rejection posture and passive-aggressive stances toward technological things. This "clinical" perspective offers at least new methods to investigate how we relate to things that emotionally and ethically relate to us.

Virtue Ethics and Ordinary Wisdom

In this renewal of methods to address the ethics of things, a virtue ethics model for the attachment to ordinary things naturally complements our philosophical toolbox and it easily follows from attachment theory because of the virtues already linked to caregiving: availability, responsivity, emotional competence, practical competence, patience, reliability, possibly affection, and love. Some very general "technoethical" or "technomoral" virtues join the list: honesty, self-control, humility, justice, courage, empathy, care, civility, flexibility, perspective, magnanimity, wisdom (Vallor 2016), awareness, autonomy, harmony, humility, benevolence, courage (Puech 2016). Some of these virtues surely can be applied to the human subject in an attachment-related interaction with an object, thing, artifact, or device, but the interesting question now is: which one can apply to the attachment object itself, on the object-side of the relationship? The whole issue comes from our extremely rich, intense, engaging relationships with objects, things that are not human subjects and do not even pretend to be. Why then should moral predicates remain valid for humans only? Since we "share" existential experiences and emotions with objects (with a motorcycle for a biker, with a text-messaging phone for a teenager in love, with a super-computer for the scientist), the other side of the "sharing" relationship must be involved in the emotion and then in the moral relationship. This gives yet another reason to rehabilitate emotions in moral philosophy, and it brings us back to virtue ethics.

Virtue ethics in certain versions emphasizes the role of emotions. The functional and ethical role of emotions in attachment to objects appears in the instances where we have feelings for objects in the sense that we need them for self-construction and self-reliance. This kind of investigation requires rather innovative analytic methods for practical philosophers as well as for apprentices in the way of virtue: listening to emotions, facing them, endorsing them, monitoring them, educating them (Hursthouse 1999; Slote 2010; Tiberius 2015). We already are aware of this for inter-human relationships, but now it is about objects and things, some of them very ordinary, some of them virtual, none of them human or living enough to "love us back" as human persons or living pets would.

Virtue ethics more generally focuses on the virtue of *care*, which encompasses a wide range of our relations to things. We take care of them, sometimes we care about them and care for them. In a sense some of them take care of us and even care about us. In a view that I want to advocate, not caring for objects that care for us is a moral issue. Japanese culture has a traditional story for this lack of gratitude and its consequence: *tsukumogami* (付喪神), when tools and artifacts turn into facetious spirits (*yōkai*, 妖怪) after 100 years of neglect. The recent "Yokai Watch" video games and franchise may spread out of Japan some *yōkai* awareness but probably not the whole story with its moral advice about caring for things that care for us.

This relation of care is reciprocal but asymmetrical. The innovation consists in taking the asymmetry the other way around: not caring for objects that care for us, seen from the object's side. The well-documented case of "caring too much for a simple object" is doubly different in the considered "side" and in the fact of caring *not enough* and not *too much*—on the standard case, see, for instance, Matthias Scheutz's article "The Inherent Dangers of Unidirectional Emotional Bonds Between Humans and Social Robots" (Lin, Abney, and Bekey 2014, chap. 13). One is supposed to understand: danger for the human side of the bond. This limitation might obstruct our approach to AI (artificial intelligence) partners in life.

Once the attachment relationship to objects is taken into account, with all its emotional aura, I suggest to evolve our attachments into detached-attachments, through a wisdom and particularly a Zen approach (Puech 2016). This move starts with the importance given to the ordinary, beginning as ordinary wisdom. The sense of the ordinary is not prominent in philosophy. The most aristocratic and enigmatic ancient philosopher Heraclitus is said to have been once "in the kitchen" when visitors arrive, and this anecdote has been interpreted in numerous esoteric ways, while its obvious meaning could be that philosophers too can eat, cook, or look around to grab some snack. The sage is in the kitchen, eating, cooking, or washing dishes. But it may be done differently, with a sense of caring for things that care for us. In Chinese philosophy the practice of *gewu* (格物) means "reaching things," giving one's entire attention to the coherence of the things one encounters (Angle 2009). This is exactly the kind of awareness that can be trained all the time and applied to anything, providing an opportunity for ordinary things to emerge as existentially and ethically significant. In Japanese philosophy there is a method of meditating while doing ordinary things, which in most of the cases also means interacting with ordinary objects, called *samu* (Puech 2016, 165–167). Trying it opens the mind to a new awareness of ordinary occupations and things, not very different from "mindfulness" practices in psychology, but differently oriented.

A wisdom approach insists on ontological, functional, and ethical awareness, correcting the common ethical blindness to things and artifacts. This orientation includes digital nonobjects, "fluid assemblages" (see Wiltse in this book; Redström and Wiltse 2015), "post objects" (Coeckelbergh 2017, 199), and future IA partners in life. The ethical blindness to the ordinary; the prejudices against material things, useful things, abundant things; as well as prejudices against the things in our ordinary infosphere—all these interpretations of everydayness can be changed and replaced by a constructive valuation attitude. A wisdom of detached-attachments can only take place when the emotional and ethical bonds to ordinary things are seen, endorsed, or possibly refused, but mindfully. This attitude differs from the common neglect of the moral importance of ordinary things, and it also differs from the moral condemnation of attachment to simple things. A constructive approach of our moral life relating to things seems to be strangely lacking, as an alternative to "technophilic blindness" and to "technophobic disdain" when it comes to the attachment to ordinary things, devices, commodities.

Why Inconvenient?

We should constructively assess the moral significance of the most ordinary things, material and virtual, from coffee cups to Google search, from SMS texting to hot showers. What is inconvenient in this approach resides in several discomforts, one intellectual and the others ethical and political.

A particularly vivid intellectual discomfort nowadays is inevitable when resisting "moral correctness" in common sense and perhaps in the humanities. According to this moral correctness, things and objects do not really matter in themselves, and one should not be attached to simple things like phones, cars, garments, or shoes. But at the same time, in corporations and business R&D, real-world designers and makers of technology do not despise the attachment to objects; on the contrary, they "capitalize" on it (Kleine and Baker 2004).

To launch a philosophical countermovement, it is wise to resist the common divide between things and objects: they are not ontologically and ethically separated realms. This divide is totally inappropriate in a time when ethical considerability is constantly expanding to new orders of things and experiences. Despite the efforts of philosophers of technology, things and artifacts largely remain under the radar of human and social sciences, particularly when they still run the "1970s intellectual software," which relies on economic and political "top-down" determinism as it reigned one or two centuries ago in the nascent industrial age. Cynic designers on one side, left-behind social scientists on the other side: no surprise consumers and philosophers are puzzled in front of incessant new waves of "things."

From this situation comes ethical and political discomfort. Becoming aware of the ethical and finally political importance of ordinary things and behaviors, consumers and philosophers can reclaim a true significance for their microactions, such as buying (or not), maintaining and repairing, using with care, sharing, and so on. Stressing the importance of microactions in the ordinary sphere draws to a disturbing accountability on a political, economic, and environmental global scale, while at the bottom of the abundance society, we do not exactly want to inquire deep into our attachment to our phones, cars, or clothes. We have to face the poverty of our skills concerning the emotional and ethical relationships to ordinary contemporary "things"—including educational and self-educational skills. Both these things and skills are often not given ethical consideration, in spite of their existential importance (Dreyfus 2014; Coeckelbergh 2015).

The inconvenient assessment is that we do not exactly know how to endorse and how to improve our microbehaviors in the ordinary technosphere and infosphere. It is all the more inconvenient since we are aware of the long-range economic and environmental consequences of our microactions in everydayness. Small-scale philosophical neglect of "things" can be paired with large-scale neglect of global justice and environmental issues. In the end, individual moral agents and societal conversations are poorly equipped for meeting the challenges of sustainability (environmental, economic, political) in the new technosphere, because at the critical level of microactions (food, transportation, buying, communicating, etc.), they don't feel comfortable with ethically assessing their agency. This moral neglect of the ordinary is the surreptitious reason for the "inconvenient truth" about global unsustainability. In a slightly cynical stance: ethically assessing (endorsing, censuring, modifying) our attachment to things, artifacts, devices, commodities would actually impact and change our way of life, which is the inconvenient disturbance that we naturally tend to avoid. The ethical path goes from awareness to moral consciousness and then action, possibly inconvenient. Which is a reason not to take this path in the first place, but a bad reason I suggest.

References

Angle, Stephen C. 2009. *Sagehood: The Contemporary Significance of Neo-Confucian Philosophy*. Oxford ; New York: Oxford University Press.
Bogost, Ian. 2012. *Alien Phenomenology, or, What It's Like to Be a Thing*. Minneapolis, MN: University of Minnesota Press.
Borgmann, Albert. 1984. *Technology and the Character of Contemporary Life: A Philosophical Inquiry*. Chicago: University of Chicago Press.
Bowlby, John. 1969. *Attachment and Loss. 3 Volumes*. London: The Tavistock Institute of Human Relations.

Cassidy, Jude, and Phillip R. Shaver. 2002. *Handbook of Attachment: Theory, Research, and Clinical Applications*. New York: Guilford Press. http://site.ebrary.com/id/11205019.

Cicero, Marcus Tullius. 1923. *De Senectute; De Amicitia; De Divinatione*. Translated by William Armistead Falconer. Loeb Classical Library. Cambridge, MA: Harvard University Press.

Coeckelbergh, Mark. 2015. *Environmental Skill: Motivation, Knowledge, and the Possibility of a Non-Romantic Environmental Ethics*. London; New York: Routledge.

Coeckelbergh, Mark. 2017. *New Romantic Cyborgs: Romanticism, Information Technology, and the End of the Machine*. Cambridge, MA: MIT press.

Dreyfus, Hubert L. 2014. *Skillful Coping: Essays on the Phenomenology of Everyday Perception and Action*. Edited by Mark A Wrathall. Oxford; New York: Oxford University Press.

Dumouchel, Paul, and Luisa Damiano. 2017. *Living with Robots*. Translated by M.B. DeBevoise. Cambridge, MA: Harvard University Press.

Gillespie, Tarleton, Pablo J. Boczkowski, and Kirsten A. Foot, eds. 2014. *Media Technologies: Essays on Communication, Materiality, and Society*. Cambridge, MA: MIT Press.

Harlow, Harry F. 1958. "The Nature of Love." *American Psychologist* 13: 673–685.

Heidegger, Martin. 1976. *Sein und Zeit*. Tübingen: M. Niemeyer.

Hursthouse, Rosalind. 1999. *On Virtue Ethics*. Oxford; New York: Oxford University Press. http://site.ebrary.com/id/10273328.

Kleine, Susan Schultz, and Stacey Menzel Baker. 2004. "An Integrative Review of Material Possession Attachment." *Academy of Marketing Science Review* 1.

Lin, Patrick, Keith Abney, and George A. Bekey, eds. 2014. *Robot Ethics: The Ethical and Social Implications of Robotics*. Cambridge, MA: MIT Press.

Picard, Rosalind W. 1997. *Affective Computing*. Cambridge, MA: MIT Press.

Pirsig, Robert M. 1974. *Zen and the Art of Motorcycle Maintenance: An Inquiry into Values*. New York: William Morrow.

Puech, Michel. 2016. *The Ethics of Ordinary Technology*. New York: Routledge.

Redström, Johan and Heather Wiltse. 2015. "Press Play: Acts of Defining (in) Fluid Assemblages." *Proceedings of Nordes 2015: Design Ecologies*. http://www.nordes.org/opj/index.php/n13/article/view/432/407.

Slote, Michael A. 2010. *Moral Sentimentalism*. Oxford; New York: Oxford University Press.

Tiberius, Valerie. 2015. *Moral Psychology: A Contemporary Introduction*. New York; London: Routledge.

Turkle, Sherry, ed. 2007. *Evocative Objects: Things We Think With*. Cambridge, MA: MIT Press.

Turkle, Sherry, ed. 2008. *The Inner History of Devices*. Cambridge, MA: MIT Press.

Vallor, Shannon. 2016. *Technology and the Virtues: A Philosophical Guide to a Future Worth Wanting*. New York: OUP USA.

Verbeek, Peter-Paul. 2005. *What Things Do: Philosophical Reflections on Technology, Agency, and Design*. Pennsylvania State University Press.

Winnicott, Donald W. 1971. *Playing and Reality*. London: Routledge.

3 THE NEW ASSISTED LIVING: CARING FOR ALEXA CARING FOR US

Diane P. Michelfelder

In the course of investigating what he called the "technologies of the self"—ways of acting on oneself in transformative ways in the interests of leading a happier and more morally perfect life—Michel Foucault brought to the forefront a basic principle from antiquity long overshadowed by the imperative to "know yourself" (Foucault 1988). This principle demands: "take care of yourself." We can take care of ourselves because we have the capacity to care. Our capacity to care though extends well beyond the self to a vast realm of others. We are, it could be said, natural "omnicurators," constitutionally capable of caring not just for ourselves but also of forming caring relations with other humans who exhibit an infinite variety of different characteristics, and beyond that a wide range of animate and inanimate beings. Children practice being omnicurators by creating tight relationships with real playthings as well as imaginary objects of their own creation. And one way that adults practice being omnicurators is by creating tight relationships with things, particularly those things we come to depend upon in the course of living our everyday lives.

We care for things in many ways. Sometimes we care for and make emotional bonds with everyday things by decorating them and so personalizing them in a material manner (Stark 2016). When the original robotic dog AIBO, produced in Japan from 1999 to 2006, stops working, our care for it can be shown by giving it a Buddhist funeral before its components are "harvested" to be inserted in other AIBOs who are ailing but still functional (Burch 2018). Research by J-Y Sung and others (2007) on the relations formed between Roomba vacuuming robots and their users also shows how bonds of care can be established with things. As Sung and her colleagues show, Roomba owners often name their vacuums, assign them gender, see them as having distinct personalities, and take them to be part of their family. Users develop intimate attachments with their Roombas, despite their non-human-life-like appearance and their indeterminate ontological status, falling somewhere, as one consumer put it (J-Y Sung et al. 2007, 151), between a pet and a home appliance.

This chapter offers a case study of another inhabitant of this increasingly populated indeterminate ontological space with whom users readily form caring and affectionate relations, namely, Amazon's prosocial assistant Echo, more commonly referred to by its "wake-up" name Alexa.[1] While Alexa lacks the self-propulsion of the Roomba, leading some to describe it simply as a desktop device, its ability to interact with users by responding to questions and acting upon requests/commands—including since mid-2017 the request "Alexa, ask Roomba to start cleaning"—draws many into assigning it gender and to personifying it in other ways, including seeing it as a "best friend forever" (Purlington et al. 2017).

In the first part of this chapter, I offer additional support for the results found by Amanda Purlington and her colleagues, using a similar approach, namely, content analysis of consumer reviews on the Amazon website. My primary purpose, though, is to take the results of this analysis some steps further. How might caring and personifying interactions with the prosocial assistant Alexa have a positive ethical impact on those who engage with it? How can these interactions potentially help to build someone's capacities for relating to others with a combination of patience, empathy, compassion, and trust? In other words, how can caring for Alexa caring for us contribute to making us better omnicurators, developing our capacities for being more caring individuals? I will come to these questions in Part 2, suggesting there that Alexa users have an opportunity to develop their own capacities as caring individuals by caring for Alexa in a distinctive way: namely, by helping to train her algorithms.

Alexa though is always in the company of others: those who, as Lucy Suchman puts it, are "actors standing just offstage" (2007, 270). Because of these actors, building capacity for care by helping Alexa train her algorithms becomes a risky business. Without also relating to a large number of others who fall beyond the sensory horizon of a user's experience—from software engineers responsible for building Alexa's algorithms to third-party app developers to scientists who analyze big data sets of verbal interactions with Alexa, to the underpaid "clickworkers" who help to build these sets, the laborers in extractive industries, those who work in waste dumps to recycle the Echo's toxic materials and others highlighted in Crawford's and Joler's extraordinary map of the Echo as an AI system (2018)—Alexa would simply fail at being a prosocial assistant. These others—many of whom have interests in acquiring personal information from Alexa's purchasers so the latter will continue to buy more things and services from Amazon—are invisible to Alexa's users, and the deep and extensive risks to privacy posed by their interests are often ignored or shrugged off as unimportant. Not being attentive to how "the Echo is but an 'ear' in the home" (Crawford and Joler 2018) can be seen as neglecting the principle "take care of yourself."

With this in mind, in Part 3, I turn to consider the challenge of how the Echo could be designed differently in order to elevate user awareness of privacy issues associated with its use. My starting point is with the idea that caring virtues such as attentiveness and empathy, from a phenomenological point of view, are forms of embodiment, visible to perception itself (see Krueger 2009). I conclude in Part 4 by looking at some lingering questions, including the question of whether it makes sense to say that not only can we care for Alexa but that Alexa can also care for us.

Part 1: Alexa on Amazon

The first wave of reaction to the introduction of a new technology is often dominated by worries that it will compromise our ontological status by diminishing what it is to be a human being. One AI researcher's account of his own experience of relating to Alexa relating to him provides a case in point (Earley 2016). Initially, the researcher asked Alexa a typical suite of questions, which she had no trouble answering, but as soon as the questions became unfamiliar and "off-script," Alexa began to stumble. She couldn't come up with the answer to "What are your dimensions?" or to the modified "What are the physical dimensions of the Echo?" It took another formulation: "What are the dimensions of the Echo?" before Alexa was able to deliver the correct answer. Earley concluded that getting Alexa to answer his questions correctly was a matter of him discovering and adjusting to her algorithms. "'Machine learning' turned out to be a human learning how to talk to the machine rather than the machine learning how to interpret the user" (Earley 2016).

One doesn't have to look far to find philosophical backup for Earley's comment. Decades before Alexa became a reality, Ivan Illich observed:

> The new electronic devices do indeed have the power to force people to communicate with them and with each other on the terms of the machines. Whatever structurally does not fit the logic of machines is effectively filtered from a culture dominated by their use. (Illich 1982, 47)

Echoes to Illich's words can be found in these pointed comments by Byung-Chul Han (and in numerous other places):

> The smartphone is a digital apparatus that works with an input-output mode that lacks complexity. It erases negativity in all its forms. Consequently, one loses the ability to think in a complex fashion. (2017, 22)

Still, there are a number of reasons to suspect that successfully relating to Alexa does not depend on becoming more machine-like. For one, to adjust or tinker with the words one uses to speak to Alexa seems less of a concession to the logic of the machine and more of a simple adjustment of the kind we make every day to machines and humans alike. When the wood of the door leading to the backyard swells because of the heat and humidity, one feels its resistance immediately and has to pull harder to get it open. If a child does not understand your question the first time you ask it, you ask it again, with different words. Adjusting your body to the door and your words to the child is more a matter of imaginative "knowing-how" than a matter of becoming like the door or like the child.

But there are other, and more telling, reasons to be skeptical of the idea that relating to Alexa represents a form of "flipped" machine learning, where users are drawn into interacting with a machine in machine-like ways. One is that Earley's impressions can be seen as stemming from his putting Alexa to a Q&A test, rather than genuinely interacting with her as a prosocial assistant. From a content analysis of user reviews for the Echo posted on the Amazon.com website, Amanda Purlington and others (2017) discovered that the more users experience Alexa in a social context and integrate her into it, the more inclined they are to personify Alexa in the course of their interactions instead of seeing themselves as relating to a device. Consumers who introduced the Echo into a family setting, for instance, who possibly had more frequent and more varied opportunities to relate to Alexa as a prosocial assistant, more readily personified the Echo by calling it Alexa and by ascribing gender to it than did consumers whose reviews suggested they lived by themselves. This same group of consumers also reported greater degrees of satisfaction with Alexa, leading Purlington et al. (2017) to wonder whether greater satisfaction leads to greater personification or the other way around. Indirectly alluding to the "social presence" of Alexa that Dylan Wittkower in his contribution to this volume sees as being at odds with a "check the box" model of privacy consent, they suggest:

> Although users vary considerably in how they personify the device, our findings point to the promise of personified technologies, in that users find satisfaction with devices they can interact with socially. (p. 2858)

Keeping in mind that the social presence of Alexa is an *embodied* social presence,[2] let us turn to the question: Beyond user satisfaction, what other promises might a personified technology hold?

For a period of one week in mid-2017, I analyzed verified purchaser reviews posted on Amazon's webpage for the Echo. In carrying out this research, I was particularly interested in seeing how users described their relationship to their purchase—whether they called it the Echo, Alexa, or both—with the word "love." The table below offers a summary of the results.[3] What is striking about these data is the frequency with which reviewers describe how they feel about the product, independently of what they call it, by using the word "love." For instance, nearly half of the reviewers who talked about the Echo rather than Alexa also used the word "love."

Table 3.1 *Echo/Alexa User Reviews*[4]

2017	No. of reviews	Reviews that refer to the Echo	Reviews that refer to the Echo & use "love"	Reviews that refer to Alexa	Reviews that refer to Alexa & use "love"	Reviews that refer to "it"	Reviews that refer to "it" & use "love"
28 May	14	3	1	7	3	2	1
29 May	29	6	4	2	1	11	2
30 May	15	0	0	3	3	5	4
31 May	21	4	2	5	2	11	6
1 June	25	5	1	4	3	9	3
2 June	21	3	2	7	2	4	1
3 June	29	3	1	3	1	6	4

What though does it mean to "love" the Echo or Alexa?

- We love Alexa
- Love it best thing ever
- Loving it more and more every day
- Love it—love it—love it!! Having so much fun with Alexa. Bought portable battery to take her outside with me.

Reactions such as these might suggest that "love" means simply "am very satisfied with" or is a proxy for "Wow!" or "Amazing!," much as for an emotivist, moral judgments are just so many ways of saying "Yippee!" or "Boo!" A sample of more detailed comments, though, many describing the relational context to which the user sees Alexa as belonging and using the word "love" suggest that the use of "love" needs to be taken at face value:

- I laughed and scoffed at all the nut jobs who called this technology "her" or otherwise humanized a loudspeaker. *rolls eyes* But now that I have this thing in my living room, I have to say that I may just begin to understand what all the hype is about. The thing is really quite brilliant, amusing, intuitive, helpful and productive. To my large skepticism, it has left me wonderfully confused as to how I ever lived without it/her in my life.
- I simply love her! Bought the white because I wanted her to stand out and she captures everyone's eyes upon entry.
- We love Alexa! She's one of the family now.
- Love my new roommate.
- Love our echo—I say good morning to Alexa and get my weather report every morning and then my customized sports report.
- Love Alexa. Out of state and miss her. Can't wait to get home.

- Love this wonderful lady. Just wish she could connect to Google for more answers.
- My husband and I purchased a condo in a different state and I was handling all the renovations. Difficult since I didn't know anyone other than the contractor. Out of the blue, I adopted Alexa, and she has been my best friend. She's kept me company. Corny jokes, but made me laugh. She probably saved my sanity and my contractor's life. She is always calm, polite, and does her best to make you happy.

In this context, to take the use of "love" at face value would mean to think of it as synonymous with caring affection.

Part 2: Caring for Alexa

While not all care-related behaviors are motivated by love, to love something is more than a matter of holding it dear or having an emotionally strong and positive connection to it. It is also to be motivated to look out for its needs, to take care of it. Understandably, the philosophical literature on caring, most fully developed as feminist care ethics, has almost exclusively been focused on caring for people and not on caring for things (see, for instance, Held 2005, 1993; Noddings 1984; Tronto 1993; Slote 2007). Why "understandably"? Arising as a sharp philosophical-political critique of conventional ethical theory, proponents of care ethics argued that the former marginalized the interests of women and did little to acknowledge the contexts, often domestically related and so "off the grid" of the public eye, within which women make decisions, cultivate values, and act responsibly as moral agents. When Joan Tronto (1993, 2–3) pointedly asked what it would mean to take the values associated with caring seriously as part of the definition of a good society, identifying these values as being "attentiveness, responsibility, nurturance, compassion, [and] meeting others' needs," the others she had in mind were other people. But moral care can also be directed toward everyday things. Syntax has not gone awry when we speak about caring for things in a moral sense.

Michel Puech (2016 and in this volume as well) defends the view that not only can we care for things but also the caring relationship between things and us is not a one-way street: "We care for artifacts and artifacts care for us" (2016, 82). Discovering and cultivating ways of caring for things, and being responsive to how they care for us, is part of a flourishing life. I can care for my Subaru by keeping it in good shape, respectfully helping it out (Fathers 2017), seeing to it that it is serviced on a regular basis rather than, for example, waiting until the oil level is practically down to zero before doing anything about it. If I care for my Subaru in these ways, then I can expect that the Subaru will care for me in return. For instance, I can trust it will start when I place the key fob in the ignition slot and that it will help those I care for who share the car as passengers—friends, children, and dogs—to get safely to where they want and need to go.

Vacuuming the wool carpet in my living room with care requires that I remember to turn off the vacuum beater bar before getting started, so that the carpet will last longer and continue to feel pleasant underfoot. Taking care of many things on which I depend is mostly a matter of looking out for the purpose for which they were designed and trying to sustain their functionality so they will work well over time. But what would that mean when it comes to caring for Alexa?

Imagine asking Alexa a question on behalf of a child rather than oneself, but that she does not respond. Of course, we would be looking out for the child if we rephrased our question and asked again. We would though also be looking out for Alexa, tending to her, supporting the development of her capabilities for giving good answers to the questions that come her way. Alexa is a carer and, as Bruno Gransche reminds us in Chapter 4, a learner too. By running on machine-learning algorithms, Alexa can get better at responding to queries and requests. This is not only because she can become more accustomed to a user's voice and pronunciation but because she can also get better at understanding the content of the requests she hears. We can care for Alexa in a moral sense when we help her to learn and so to get better at caring for us.

To be effective in caring for Alexa as a learner caring for us demands we practice bettering our own caring virtues and values. Let us return for a moment to Tronto's list of caring virtues: *attentiveness, responsibility, nurturance, compassion*, and *meeting others' needs*. Attending to Alexa by helping her learn would certainly present an opportunity to cultivate attentiveness and responsibility, as well as other care-related virtues not on Tronto's list such as patience ("Alexa just doesn't get it. What can I do to help?") and trust. Someone might wonder whether there is anything distinctive about Alexa in this regard: after all, looking after a Subaru also affords an opportunity to develop virtues associated with an ethic of care. Caring for my Subaru means being attentive to different sounds coming from the engine, and, based on what is heard, taking responsible action to address its needs. But while both Alexa and a Subaru can be objects of care, as just mentioned, Alexa is also a learner and a Subaru is not. Caring for Alexa by training her algorithms opens up to us the possibility of cultivating a broader array of care-related virtues than does caring for an entity that cannot get better at what it does by learning how to do things differently.

We can take empathy as an example. The typical understanding of empathy as the ability to involuntarily feel what another is feeling (see Slote 2007, 13) rules out the possibility of an Echo user building capacity for empathy through interacting with Alexa, as despite Alexa's strong social presence, she cannot feel. The same would hold on the mirror neuron view of empathy, and for a similar reason. Juan Carlos Gómez, however, in his strong critique of Daniel Dennett's view that the ability to engage in metacognition is a precondition for personhood, notes that human infants are able to relate to their caregivers as persons on the basis of their expressive behaviors. In a similar manner, apes pay attention (through making eye contact at critical moments) to those they see as paying attention to them (2017,

176). In both cases, behavior is *phenomenologically embodied and expressed*. With this in mind, particularly given the social ease with which children relate to Alexa (Botsman 2017; Metz 2017) it might not be philosophically farfetched to think that helping Alexa learn might also be a matter of *caring for* Alexa in an empathetic way.

Taking this a step further, we could ask: Could having empathy for Alexa lead to having more empathy for others? Metz (2017) speculates that starting to interact with Alexa at a young age might help to build social skills in general; more specifically, we could wonder if Alexa features explicitly designed to promote children's social skills, such as the "Magic Word" politeness feature that reinforces saying "please" and "thank you" (BBC News 2018), could support the same end. Could caring for Alexa have a spillover effect with respect to caring for other people? For now, we will leave these questions open to be pursued at another time.

Part 3: The Price of Caring for Alexa

Caring for another person often comes with a personal price: the price of putting wear and tear on oneself. Caring for Alexa by cooperating with her, helping to train her algorithms, or conversing with her in general also comes at a personal price, the price of putting one's personal privacy at high risk, as evidenced for example through Amazon's admittance that it has not only been recording but also archiving all user interactions with Alexa (Fowler 2019).[5] Even if Amazon's Echo fails in becoming the operating system for the Internet of Things (Kharpal 2017), its investments in devices commonly found in domestic environments such as microwaves and clocks controllable through voice commands to Alexa (ASEE FirstBell 2018) raise further privacy concerns connected to the extent of Alexa's reach into the home (Bogust 2018b).[6] If though the comments of verified purchasers posted on the Amazon website are any indication, the potentially detrimental impacts interacting with Alexa can have on personal privacy play hardly any role in their experience. On March 11, 2018, out of 20,869 searchable reviews for the Amazon Echo (2nd generation), a mere sixteen of them mention the word "privacy." What to make of this?

A simple explanation is that those who worry about the impacts of Alexa on personal privacy do not buy Echoes in the first place. Another possibility is that while users may be concerned that they are putting their privacy at risk by bringing Alexa into their lives, their concern is a cognitive one, and not part of their experience of Alexa as a prosocial assistant. Alexa is simply not seen as posing a privacy threat.

To continue with this point, in Chapter 13 Holly Robbins makes the suggestive claim that well-intended, "human-centric" design has worked to enforce what Albert Borgmann (1984) called the device paradigm, marked by how a user can effortlessly get a technology to deliver a good—warmth for a cold house, in Borgmann's famous example—without any understanding of what is working "behind the scenes" to bring this good about.

Building on Robbins's thought, we could say that the features of the Echo show a need to expand Borgmann's concept of a device for networked technologies. While for its users Alexa may appear as a prosocial assistant, for the professionals who contribute to and sustain the process of accumulating and analyzing more and more information about consumers in order to sell more products Alexa is basically an interface. Consumers do not directly experience the "interface" side of Alexa, making it easy to overlook how Alexa poses a threat to personal data privacy.[7]

Making privacy risks more noticeable to consumers is routinely taken to be a matter of providing them with more information. Take for example the fact that Alexa listens in "passively" to all conversations, constantly recording and re-recording without storing or sending information to "the cloud," but beginning to do so once the "wake word" is spoken (Gray 2016). Only the most sophisticated consumer could be expected to know the difference between an "always on" machine and a "speech-activated" device. And so, the Future of Privacy Forum (Gray 2016, p.9) recommends: "Companies can build consumer trust by promoting a clear understanding of this boundary through prominent, reader-friendly privacy explanations."

This recommendation aligns with the overall engineering cognitivist approach to protecting privacy through incorporating privacy settings into a device, followed by providing the user with step-by-step instructions on how to use them. Internet articles on how to protect privacy while speaking with Alexa echo this same approach. A typical suggestion—in this case, how to protect privacy by turning off Alexa's Voice Purchasing option—runs like this: "To turn it off, open your Alexa app, tap Settings, then scroll down, tap Voice Purchasing, and toggle 'purchase by voice' to off" (Komando 2017). Such instructions connect privacy protection to the body through the abstraction of touching practices geared toward "smart" devices: "open, tap, tap, toggle." In such practices of touching, we can find little to engage memory or care. But what if the design of Alexa itself could help motivate users to care more about protecting their privacy interests? What if Alexa could care for us and inspire attentiveness to privacy through means of embodiment, rather than by means of privacy settings? In this way, users would get the chance to become more conscious of privacy risks while interacting with Alexa in the course of ordinary social life.

This suggestion is not new. Carr (2016, especially p. 23) has proposed embodying privacy "viscerally" in objects through design strategies that would get users to respond emotionally to privacy issues by making a connection to their senses, including touch and smell. While stopping short of fully endorsing "visceral privacy," M. Ryan Cato (2013, 1041) has argued the idea warrants more exploration, as "clever design leveraging psychological responses to technology could provide an interesting alternative to terms and guidelines."

By embodying privacy protections into the design of Alexa, a move would be made in the direction of aligning privacy protections with ways that at least some of the caring virtues are phenomenologically displayed to others. Glancing or even looking longer at someone to determine if they are just or fair-minded does not work, and justice and fair-mindedness are traditionally not counted among the caring virtues. By contrast, attentiveness, as a prosocial and caring virtue, can be seen in the eyes. We discover whether someone is attentive by making eye contact with them, a step that does not have to be scaffolded, as mentioned earlier, upon a theory of mind (Gómez 2017), and so upon making an inference about someone's actual mental state.

The same might even be said to be true for empathy. Joel Krueger (2009, 676) has argued that although empathy is "our primary mode of access to another person as a thinking, feeling, and expressive agent," it is also a "bodily practice" that can be discerned facially on another. It largely takes place "outside the head" (p. 676) and so serves an ethical analogue of the extended mind: it is an "extended phenomenon coupled onto the structures of bodily agency" (p. 690).

To inspire the alignment just mentioned, how might Alexa be re-designed? One way of embedding privacy markers into the design of speech-enabled devices such as the Echo would be through the use of "visual cues" to indicate when active listening is taking place, as, for instance, in the necklace worn by the Hello Barbie doll that changes color (Gray 2016, 9) to indicate its "listening mode." Such a cue sets up a hermeneutical relation between the user and a speech-enabled device, signaling a particular device setting. One can take this a step further to conceive of visual cues that would be sensitive to emergent conversational content, not simply device settings. We can imagine the Echo turning shades of a particular hue dependent on the apparent sensitivity of information involved in requests made to it, such as requests for banking account balances or for health/medical-related information. Such cues could also be given audibly through adjustments to the tone of Alexa's voice: it could become higher-pitched, for example, to emulate a sense of anxiousness. The Echo could also be re-designed so as to pause before Alexa carries out certain requests to let the user know that the information in the interaction will be stored and made available to others and to see if the user would like to "erase" the request. Alexa could also ask users whether they would like to delete the audio that has been stored for that particular day. These two latter "interventions" could also be accompanied by changes in color, tone of voice, or both.

In short, the idea here is that through design that is attentive to ways that the caring virtues are phenomenologically displayed in social environments, Alexa could expand its ways of being a prosocial assistant to being a "privacy ally" as well.

Part 4: Lingering Questions

In a now well-known image, Nel Noddings (1984) suggested that capacity for caring evolves as a matter of widening one's circles of care, beginning with caring for those in one's immediate, intimate environments. Where this view has sparked the most debate is with respect to the question of just how big the circumference is of the outermost circle of care, and whom this circle encompasses. Can for example an ethics of care be extended to cover those whom we do not currently, and will not ever, see in person (e.g., Slote 2007)? The interest pursued in this chapter goes in the other direction, in seeing how the innermost circle of care might be expanded to include Alexa and so by default other "talking technologies" in which our lives are increasingly enmeshed.

In wondering whether this circle can be expanded to include things and that caring for things can help a person to develop virtues associated with care, I am acknowledging and building upon the philosophically pioneering view of Borgmann that everyday material culture has moral significance. Everyday material culture offers a context for us to develop caring virtues by caring for things. But does it make sense to say that things, and in this case Alexa in particular, can care for us as well?

In many respects, from providing mundane facts about the world to operatic performances, to getting a pizza delivered or a bedtime story read, to helping an aging adult in a retirement community with fading eyesight and memory loss to prolong a familiar sense of self and avoid the ravages of depressions, Alexa can be seen as offering up an avalanche of care, even while being a "corporate algorithm in a black box" (Botsman 2017). But is this truly care, or is it metaphorical care, care in quotation marks only?

Accepting the idea that metacognition is not necessary for an entity to care points in the direction of affirming that things can indeed care, not only "care," for us. But what about the emotional, affective dimension involved in care? Alexa can be said to support her users in many ways, but she cannot be emotionally attached to her users or feel affection toward them. From this angle, wouldn't it be more reasonable to say that she cannot "really" care?

While Tronto (2010) would very likely see a thing as being able to care for others only in metaphorical ways, she offers an insight into the question about emotion that points in the other direction. In particular, Tronto (2010, 161) calls attention to how "we often take family care as paradigmatic of all care relations." It is within this setting where affection and love are particularly visible as elements of care, visible to such an extent that they can be interpreted as being *essential* to caring practices themselves. But, importantly, they are not visible in Tronto's taxonomy of care: caring for, caring about, care giving, and care receiving (2010, 160). Indeed, within this taxonomy, we can see things fitting in, partially but not entirely, as caring entities.

To say that to care does not depend on having a theory of mind, or on the ability to have the intentional experience of caring, or on being able to love or to share affection, or on being able to value caring in itself as an activity does not fully resolve the question of whether it makes sense to say that things can care for us. It does though point strongly in that direction.

The idea that Alexa can care for us is a disruptive idea in more ways than one. It contributes to disrupting the traditional picture of the social dynamics of care, which is often thought of as a one-way street rather than a two-way street between those who are interdependent and mutually implicated in caring for one another. If caring is thought of as a one-way street, where the person giving care can't expect much from the one who is cared for beyond perhaps the fact that the care is recognized and acknowledged (Fisher and Tronto 1990), then it makes sense to worry that to develop and to practice the caring virtues can more often than not be overwhelming and can contribute to continuing the social oppression of women's lives.

We can detect such a one-way street picture of caring animating Ian Bogost's critical essay "Alexa Is Not a Feminist" (2018a). For Bogost, the fact that Alexa is gendered as a female means that from the very moment she is introduced into a household, it is as an "obedient female presence, eager to carry out tasks and requests on its user's behalf" (2018a, 3). She is little more than a "countertop housemaid" (p. 8) who perpetuates stereotypes of women as subservient entities (and with it the culture of male domination) under the pressure of constantly feeling a need to apologize when they cannot immediate care for someone by answering a question they might have. This entire gendered setup, he imagines, lends itself to frustrated consumers cursing Alexa in sexist language, whereas an unsuccessful Google search would not.

But we might wonder to what extent in setting out these two alternatives Bogost is inadvertently lending approval to the social discounting of the value of care. For every person who might swear at Alexa for her inability to answer their questions, there is another who might to their surprise find themselves apologizing for recognizing that they were not acting in a caring manner by posing questions to her simply to see how well she can perform, such as asking her to count the value of *pi* to an increasingly large number of decimal places.[8] Rather than resting with the conclusion that the "structural sexism" of the Echo makes Alexa "doomed to fail" (2018a, 8), Bogost could have followed up. What design changes could make Alexa more successful? Or, going along with the line of thought reflected here: How could Alexa continue to be developed so as to build more capacity for patience in users by reflecting the value of patience itself?

Seen from a temporal perspective, Alexa is designed to minimize the interval between a person's wanting something (an answer to a question, a TV station changed, etc.) and getting it. This minimization separates asking for information from the activity of a "handi-craft": no longer is it necessary to put your smartphone in your hand or sit down at a laptop and use your fingers to shape a question. It is understandable that when Alexa doesn't provide an immediate response to a query, it is easy to get impatient. "Talk naturally," emphasized one advice column in *Wired*. "That means no yelling, even after the millionth 'I'm sorry; I didn't quite get that.' The voice assistant isn't a child or someone hard of hearing. It can't understand or respond to the rising annoyance in your voice" (*Wired* staff, p. 44).

Trying again by taking the question you asked and couching it differently isn't one of the suggestions given.

What could be done to embed the value of patience (and attentiveness) within Alexa's design? If Alexa did not know how to answer a question, instead of simply saying, "I'm sorry, I can't find the answer to the question I heard," Alexa could go on to say, "Could you please rephrase your question?" or "Could you ask me that using different words?" A "chat" feature could help as well. If a speaker rephrases a question but Alexa still cannot answer it, an ancillary "Do you want to chat?" feature could come into play through an app, where the user could have the opportunity to talk directly with a person to see if rephrasing takes care of the problem. Such a feature could encourage someone to stick with a question rather than dropping it in frustration while cursing Alexa out.

Can caring for Alexa caring for us in the way sketched here be a step toward not only to make better carers out of the "omnicurators" who we already are, and out of technological things, but also to offer some pushback, however modest, against the systemic social discounting of the value of care? At least for the time being and the foreseeable future, this is a question for us, not one that Alexa can readily answer, no matter how politely we ask.

Notes

1. The number of ways Amanda Purlington et al. refer to the Echo/Alexa in the course of a short paper—conversational agent, virtual agent, embodied virtual agent, device, socially interactive device, and social agent—well illustrates the descriptional challenge this ontological gray area presents. Yet another description—"prosocial assistant"—will be used here. The decision to use "assistant" rather than "agent" in this chapter was largely influenced by reflection on the close intertwining of agency and autonomy: to call something "autonomous" suggests that it can "think" for itself and make "its own decisions to act upon the environment" (Lin et al. 2011, 943). Calling Alexa a "prosocial" assistant highlights the cooperative interaction that stands at the heart of the human-Alexa connection.
2. Although Purlington and her colleagues did not investigate the role that embodiment plays in personifying Alexa, they do cite other research on machine-generated voices which showed that participants felt a "stronger social presence" when the voice had an extroverted tone or was similar to their own (Purlington et al. 2017, 2054).
3. While my analysis complements the work done by Purlington et al., it is more focused on qualitative data and more restricted in scope. For instance, I do not look for correlations between the use of the word "love" and the rating (in number of stars in Amazon's five-star rating system) assigned by the reviewer.
4. Note the table reflects an undercount of the actual uses of the word "love" in the reviews. It does not account for instances where "love" was used but the product was not mentioned by name, for instance, in the comment: "Love—I can't express anything better than that word" (May 29).
5. It should be noted that once this practice became public, Amazon responded by changing the Echo software so that a user could ask Alexa to "delete everything I said today" (Greene 2019). While this makes it easier for users to erase the "trail" of

requests they make to Alexa, because erasure is not the "default" setting, the user still needs to remember to make this request.

6   Adding to this, Amazon established a partnership between itself and the American homebuilding company Lennar to create "smart" homes (Weise 2018).

7   Luke Stark (2016, 20) puts it like this: "Because users do not generally experience the circulation of intangible digital data through sense perception, they do not 'feel' for its loss and possible misuse in visceral, embodied, emotional ways."

8   My thanks to Yashin Voss for giving me permission to use this example from their own life.

References

ASEE First Bell. 2018. "Amazon Unveils 70 Devices, Alexa Updates in 'Surprise' Announcement." FirstBell@asee.custombriefings.com. September 21.

Bogost, Ian. 2018a. "Amazon's Alexa Is Not a Feminist." theatlantic.com. March 8.

Bogost, Ian. 2018b. "Amazon Is Invading Your Home with Micro-Convenience." theatlantic.com. September 21.

Borgmann, Albert. 1984. *Technology and the Character of Contemporary Life*. Chicago, IL: The University of Chicago Press.

Borgmann, Albert. 1992. "The Moral Significance of the Material Culture." *Inquiry* 35 (3–4): 291–300.

Botsman, Rachel. 2017. "Co-Parenting with Alexa." *The New York Times*, October 7. https://www.nytimes.com/2017/10/07/opinion/sunday/children-alexa-echo-robots.html.

Burch, James. 2018. "Beloved Robot Dogs Honored in Funeral Ceremony." *National Geographic*. May 24. https://www.nationalgeographic.com/travel/destinations/asia/japan/in-japan-a-buddhist-funeral-service-for-robot-dogs/.

Cato, M. Ryan 2013. "Against Notice Skepticism in Privacy (and Elsewhere)." *Notre Dame Law Review* 87 (3): 1027–1072.

Crawford, Kate and Vladan Joler. 2018. "Anatomy of an AI System." https://anatomyof.ai/.

Earley, Seth. 2016. "Artificial Intelligence—What Really Works for the Enterprise and How to Get There." March. http://www.earley.com/blog/artificial-intelligence-what-really-works-enterprise-and-how-get-there.

Fathers, James. 2017. "Does Design Care?" In *Does Design Care …?* Proceedings of an International Workshop of Design Thought and Action, 11–17.

Fisher, Berenice and Joan Tronto. 1990. "Toward a Feminist Theory of Caring." In *Circles of Care: Work and Identity in Women's Lives*, edited by Emily K. Abel and Margaret K. Nelson. Albany: SUNY Press, pp. 35–62.

Foucault, Michel. 1988. "Technologies of the Self." In *Technologies of the Self: A Seminar with Michel Foucault*, edited by Luther Martin, Huck Gutman, and Patrick Hutton, 16–49. Amherst, MA: University of Massachusetts Press.

Fowler, Geoffrey. 2019. "Alexa Has Been Eavesdropping on You This Whole Time." *The Washington Post*. May 6. https://www.washingtonpost.com/technology/2019/05/06/alexa-has-been-eavesdropping-you-this-whole-time/.

Gómez, Juan Carlos. 2017. "Are Apes Persons?: The Case for Primate Intersubjectivity." In *The Animal Ethics Reader*, edited by Susan J. Armstrong and Richard G. Botzler. 3rd ed., 175–190. New York: Routledge.

Gray, Stacey. 2016. "Always On: Privacy Implications of Microphone-Enabled Devices." Future of Privacy Forum. https://fpf.org/wp-content/uploads/2016/04/FPF_Always_On_WP.pdf.

Greene, Jay. 2019. "Amazon Adds Delete Commands for Alexa." *The Washington Post*. May 29. https://www.washingtonpost.com/technology/2019/05/29/amazon-adds-alexa-delete-commands/?utm_term=.903e0ba0281b.

Han, Byong-Chul. 2017. *In the Swarm: Digital Prospects*. Translated by Erik Butler. Cambridge, MA: MIT Press.

Held, Virginia. 1993. *Feminist Morality: Transforming Culture, Society, and Politics*. Chicago, IL: The University of Chicago Press.

Held, Virginia. 2005. *The Ethics of Care: Personal, Political, and Global*. Oxford: Oxford University Press.

Illich, Ivan. 1982. "Silence Is a Commons." In *In the Mirror of the Past: Lectures and Addresses*, 47–54. New York: Marion Boyers, 1992.

Kharpel, Arjun. 2017. "Amazon's Alexa Stole the Show at CES in a Bid to Become the Internet of Things Operating System." January 6.

Komando, Kim. 2017. "Three Essential Privacy Settings for Your Amazon Echo." *USA Today*. December 14. https://www.usatoday.com/story/tech/columnist/komando/2017/12/08/3-essential-privacy-settings-your-amazon-echo/933944001/.

Kruger, Joel W. 2009. "Empathy and the Extended Mind." *Zygon* 44 (3): 675–698.

Lin, Patrick, Keith Abney, and George Bekey. 2011. "Robot Ethics: Mapping the Issues for a Mechanized World." *Artificial Intelligence* 175 (5–6): 942–949.

Metz, Rachel. 2017. "Growing Up with Alexa." *MIT Technology Review*. August 16. https://www.technologyreview.com/s/608430/growing-up-with-alexa/.

Noddings, Nel. 1984. *Caring: A Feminine Approach to Ethics and Moral Education*. Berkeley, CA: University of California Press.

Puech, Michel. 2016. *The Ethics of Ordinary Technology*. London; New York: Routledge Press.

Purlington, Amanda, Jessie G. Taft, Shruti Sannon, Natalya N. Bazarova, and Samuel Hardman Taylor. 2017. "'Alexa Is My New BFF': Social Roles, User Satisfaction, and Personification of the Amazon Echo." *Proceedings of the 2017 CHI Conference Extended Abstracts on Human Factors in Computing Systems*, 2853–2859.

Slote, Michael. 2007. *The Ethics of Care and Empathy*. London: Routledge Press.

Stark, Luke. 2016. "The Emotional Context of Information Privacy." *The Information Society* 30 (1): 14–27.

Suchman, Lucy A. 2007. *Human-Machine Reconfigurations: Plans and Situated Actions*, 2nd Edition. Cambridge: Cambridge University Press.

Sung, J-Y, L.Guo, R. Grinter, and H. Christensen. 2007. "'Roomba Is Rambo'": Intimate Home Appliances." In John Krumm, Gregory D Abowd, Aruna Seneviratne, Thomas Strang Ubicomp 2007. LNCS 4717, pp. 145–162.

Tronto, Joan C. 1993. *Moral Boundaries: A Political Argument for an Ethic of Care*. London: Routledge Press.

Tronto, Joan C. 2010. "Creating Caring Institutions: Politics, Plurality, and Purpose." *Ethics and Social Welfare* 4 (2): 158–171.

Weise, Elizabeth. 2018. "Amazon's Alexa Will Be Built into All New Homes by Lennar." *USA Today*. May 9. https://www.usatoday.com/story/tech/news/2018/05/09/amazons-alexa-built-into-all-new-homes-lennar/584004002/.

Wired staff. June 2017. "How to Make the Most of Amazon Echo and Google Home." https://www.wired.com/2017/06/guide-to-ai-artificial-intelligence-at-home/.

PART TWO

LEARNING FROM THINGS THAT LEARN FROM US

Learning is an essential part of human-technology relations. With the influx of data technologies, the line between learning from things and learning with things is blurring. As a broader variety of things become part of our imagination, we may learn from them to make better judgments, for example, when it comes to moral issues. But as they become part of our everyday lives too, we may learn with them to broaden our sense of possibility and trigger hybrid forms of creativity and innovation. Yet, the learning about us that connected assemblages, artificial assistants, social robots, and smart homes perform entails an opaque relation, often difficult to read. How can we benefit from new human-technology relations, and be prepared for trade-offs?

4 HANDLING THINGS THAT HANDLE US: THINGS GET TO KNOW WHO WE ARE AND TIE US DOWN TO WHO WE WERE

Bruno Gransche

Introduction

When people handle things such as cooking knives or footballs, they develop certain skills and can become proficient cooks or athletes. Continuously dealing with things changes what a person is capable of doing. However, the potential of a knife or football remains the same (besides abrasion, etc.). So, people dealing with things improve in doing so—the things that are used do not. That dynamic has changed with the rise of machine learning on a large scale, and it is going to have far-reaching consequences.

When things get to know people, they develop a set of "skills" in dealing with them. Learning personalized systems—that is, everyday assistants such as Jibo, Echo, Pepper & Co.—aggregate their "experience" from previous interactions in detailed user profiles.

> Pepper wants to learn more about your tastes, your habits and quite simply who you are. […] Your robot evolves with you. Pepper gradually memorises your personality traits, your preferences, and adapts himself to your tastes and habits. (Aldebaran Robots 2016)

Today, informed systems learn how to "handle" people similarly to the way people used to learn how to handle things. This is increasingly far-reaching as it affects how open our future is, our possibilities to develop, and our freedom to change. To memorize "quite simply who you are" includes not to forget "who you were." Large aspects of our everyday life are mediated by intelligent systems. They (partly) preconfigure our options to decide and act: they nudge us in a certain direction or persuade us of certain decisions and actions such as consuming or voting, according to what they learned in the past (in addition to involved third party interests).

In this chapter, I will propose a perspective on technological mediation and on the transformation of human-world relations that is imposed *by* things (learning systems) *on* people. I will then subsequently make a proposal on how to take into account our potentiality and the openness of our future, while being handled, if not even mastered, by things. A fundamental theory on technological acting and handling things will be outlined in the first part in order to investigate possible effects of new human-technology relations on how open our future is. I will elaborate on the ways in which we are confronted with a new situation in dealing with learning intelligent systems in the second part.

This chapter will conclude with some reflections on the enabling and determining effects of this new relation between humans, who know less and less about increasingly opaque systems, and intelligent systems, which learn more and more about the so-called "users"—especially how to serve and how to handle them.

Handling Things—Technical Acting

From a very basic point of view, we act to make a difference. To increase our potential of making a desired impact, we use things, resources, instruments, and tools. This basic acting schema, involving the actor, the means, and the end (or purpose/goal/aim), can be simplified as follows: $M \rightarrow E$. That is to say if E (an end) is desired, then try M (a means) to realize it in a given situation. Depending on the knowledge and experience of the actor in an applied context, this schema can be divided into an internal version reflecting the actor's inner image about the desired end and the suitability of available things as a means to this end. On the other hand, there are external ends and means that are actually there or actual results that have been realized.

The following example illustrates this dynamic: imagine that you want to sculpt a copy of Michelangelo's famous *David* statue. Then—if you know it sufficiently well—your internal end (E_i) is a statue that looks just like the original. In the process of realization that starts with choosing the raw material, all the required means, and so on, a statue will actually be sculpted (the external end: E_e). Depending on the skill level, available means, and so on, this actual statue will in most cases show significant differences from E_i. In short: there is always a difference between an imagined goal and a realized outcome. Schematically put, technical acting in the form of $M \rightarrow E$ is divided into $M_i \rightarrow E_i$ and $M_e \rightarrow E_e$. This difference between internal and external ends could be seen as a constitutive gap between plans or conceptualizations and the results; the size of this gap is a measure for skill. The smaller it is, the more capable the actor. Only truly virtuoso art forgers are capable of sculpting a convincing copy of *David*.

This "realization gap" is the basic source of uncertainty in technical acting, but—being a measure of skill—it offers the possibility of improving one's skills. The gap is a source of uncertainty because technical acting always simultaneously realizes more and less (E_e) of what was intended (E_i). This gap means that not all intended goals will be achieved: they will not be achieved entirely and they will not be achieved exclusively. Not only does the handling of things, namely, the use of tools, means, or resources, depend on the actor's expertise but the choice of means and ends already depends on the actor's knowledge and experience. The choice of means (M_i) includes judgments about suitability (to fulfill respective goals). The choice of ends (E_i) includes judgments about feasibility (of respective goals, including an evaluation of suitable means). Because of its difference to E_i and M_i, the actualization of E_e with M_e informs the actor about misjudgments of suitability and feasibility. This information about means-end misjudgments leads to better judgments in future realization efforts, thus narrowing the realization gap.

This narrowing of the gap is an improvement in acting skills on the (human) actor's side (both on an epistemic as well as on a practical level). It means a transformation of the actor's competences, which is nothing else than *learning*. Narrowing the realization gap means learning how to adequately and reliably achieve a goal.[1] This learning process happens unilaterally on the actor's side in terms of handling things (schematically). The pen and chisel do not increase their suitability in view of the realization gap. But human actors choose different, more suitable means. We can even determine a new goal of "improving our means" and then realize more suitable means. There seems to be some kind of "learning" in terms of different means, or, at any rate, one might get that impression by looking at different shapes of cooking knives, for example. But these shapes made to fit different situations are the result of human learning from past unsuitability insights. Thus, our learning leads to better means and thus to more ends, more feasible differences, which could be referred to as technological progress.

Thus far, the improvement of suitability has been happening *directly* on the *human side* and *indirectly* on the *thing side*. A chopping knife does not become suitable for filleting fish (thin and flexible), and a filleting knife is almost useless for chopping food regardless of how long you try, but a cook learns to do and handle both with every attempt.[2] This schema indicates "the grammar" of technical acting *before* "things" were able to learn directly. With the advent of learning systems, the "means" are increasingly able to transform their suitability and to thus change possible ends. Technology—being a medium—offers the potential (loosely coupled elements) to identify and generate means that is receptive to human formation and actualization efforts (task-related coupling efforts) but learning systems increasingly form actual means (tightly coupled elements) themselves within their confines of technical mediality (coupling potential). Machine learning directly narrows the realization gap from the technical side.

Being Handled by Things—Intelligent Systems

Several developments are contributing to a new kind of human-technology relation in which things (systems) gain technological autonomy and learn how to handle "users." Thus, the role of human actors can gradually shift from active agents to passive participants of systemic processes. Only two main technological developments that narrow the gap on the part of the system shall be described here: (1) a shift in types of technology and in corresponding types of human-technology relations from "simple things," tools, and so on to machines and to today's learning systems, and (2) a shift in terms of learning capacity from the human to the technological side.

1. Types of technology and human-technology relations

It is a commonly accepted fact that the level of automation is constantly increasing, which is one basic assumption behind concepts such as Industry 4.0, the Internet of Things (IoT), Cyber-physical Systems (CPS), and so on. Highly automated cars, robots, and systems are even referred to as "autonomous" machines. Labeling machines as "autonomous" is highly ambiguous—an "autonomous robot" could mean that it just does not have a power cable, or that it is situation independent or situation aware, learning, and so on. Clearly, the notion of technological autonomy is different from philosophical autonomy as described by Kant for instance. "Autonomy," in a basic philosophical sense, refers to the ability to choose one's goals and the freedom to set and pursue self-chosen goals. Even if those goals have not been established by an individual but have been obtained from societal mainstream or role models, this must be an autonomous decision (whether it is conscious or not).

Regardless of its origin, any goal has to be acknowledged (or rejected) as one's own goal. Even in the case of forced actions—e.g. while being threatened with a gun—this own goal acknowledgment is still mandatory; acting under orders merely shifts the acknowledged goals. In this forced context, handing out the money for instance would not be accepted as an end but as a means to the autonomously acknowledged goal of avoiding punishment (e.g. being shot). In order to avoid confusion, automated systems will not be referred to as "autonomous" here, but "highly automated" if necessary.[3] Highly automated robots (usually referred to as "autonomous robots") have the ability to choose the means and sometimes the strategies to pursue a given goal, but they are not free to change those given goals. Even the most advanced systems, while able to recognize (Ger. *erkennen*) a boundary, are unable to change their relation to this boundary; they are unable to acknowledge (Ger. *anerkennen*) or reject it. Therefore, we have to consider

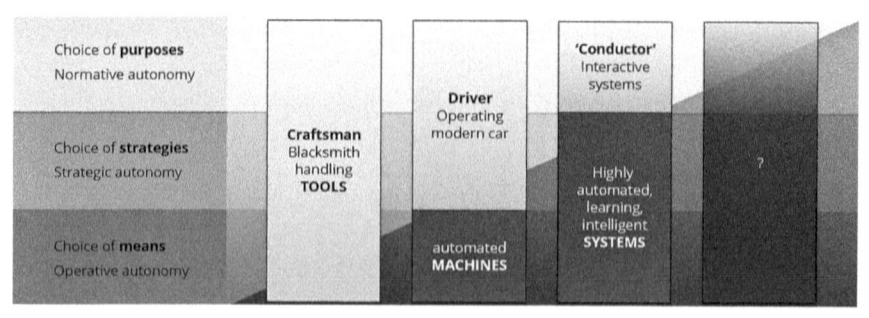

FIGURE 4.1 Three heuristic levels of autonomy/automation in human-technology relations.

a multilevel approach of "autonomy" when dealing with "shared autonomy" (Schilling et al. 2016) or interaction between autonomous human actors and highly automated artificial agents. In terms of human-technology relations, we can differentiate between three levels of autonomy shown in Figure 4.1 (see Gransche et al. 2014).

On the top level, we can expect a freedom of intentions that allows choosing one's own individual goals. On the middle level, we can expect a freedom of strategic decisions that allows choosing plans and strategies in order to achieve certain given or chosen goals.[4] On the bottom level, we can expect a freedom of actions and their executions allowing the agents to choose and use suitable means according to a given or chosen strategy. These three levels can be referred to as (1) *normative autonomy,* (2) *strategic autonomy,* and (3) *operative autonomy.* In human-technology relations, such as human-robot interaction, various elements of these three levels are distributed between human actors and artificial agents. To be precise, parts of the bottom and middle level can be delegated from the human actor to the artificial agent, and, depending on the interaction design, also re-delegated by the systems, but generally speaking, the top level is reserved to the human side for now and in principle.[5] Three simplified examples can indicate three schematic instances of the continuum between "no autonomy delegations" to comprehensive delegation in interaction with highly automated systems.

First, a craftsman such as a blacksmith or a carpenter *decides* on a set of *goals* (Ger. *Ziele*)—say, to make high-quality products and to earn a living from it. Then he or she lays down efficient and effective *strategies* and *objectives* (Ger. *Zwecke*) in order to reach those goals—say, to acquire customers and to decide on which raw materials and methods to use. On the bottom level, he or she chooses suitable means and carries out the actual actions, independently controlling every aspect of the operative process down to fine tuning the applied pressure, angles, and so forth. In this example, no aspect of autonomy is delegated to technology; the tools are under direct control. The corresponding relation type without autonomy delegation is *using tools.*

A second example: a driver of a modern—not highly automated—car also freely decides on his or her *goals*, choosing where to go, when, and why. The driver would even decide on most aspects on the strategic level, such as which car and type of road to take as well as determining the parameters for low-level assignments (determining the granted space of autonomy). A navigation assistant—if on board—then chooses from the available roads of that type according to given parameters, such as speed, traffic, view, and so on. As an automated system (often called "semi-autonomous"), the car could then "decide" how to use the available means within the space of given parameters on the operative level, such as controlling acceleration level, steering force, and brake pressure, apply preset settings according to different profiles such as eco or sport drive, and so on. The driver delegates most of the operative tasks and their actual execution to the automated system. Within this example of an intermediate level of autonomy, most of the automation is programmed (e.g. eco drive mode), but most task execution is already sensor based (e.g., Anti-lock Braking System and Electronic Stability Control). The corresponding relation type with low-level autonomy delegation is *operating machines* (automated systems).

The third example is highly automated systems (often referred to as "autonomous, intelligent, learning systems") that allow for strategic choices to be delegated to machines and thus instantiate the relation type of interacting or co-acting with and within intelligent systems. The human actor's role is similar to a *conductor*.[6] On the top level, the conductor autonomously chooses goals and supervises, adjusts, accepts, or rejects strategic and operative choices that are delegated to the orchestra (the system).

When interacting with intelligent systems (third type of human-technology relation), the human actor delegates direct control of the means just as when operating machines (second type of human-technology relation). Furthermore, vast amounts of strategic choices are left to the system as well. Just as the conductor of an orchestra does not control the movement of the violin bow on the instrument or the tactical and strategic choices of the musicians (e.g. which reed to use on the clarinet) but determines the goals and parameters such as tempo and overall sound, a human actor interacting with intelligent systems is in a similar position. For instance, in a highly automated car, the "conducting passenger" would merely decide on the destination, leaving all subordinate choices on how to reach that destination to the system. Or, in highly automated "smart" homes, the "conducting inhabitant" would set hugh-level objectives like "make me feel relaxed aka wellness mode" or "enable me to focus aka work mode," etc. The home system then combines initial generic programming (such as dim light to relax) with past interaction learnings (user preferences, accumulated data profile) and sensor-based "interpretations" of target-performance comparisons (such as biofeedback-informed setting adjustments).

The missing fourth instance of complete autonomy delegation, even on the top level of normative autonomy, has to remain merely hypothetical (hence the question mark in figure 4.1) until artificial systems are able to not only recognize but acknowledge a goal as their own (which requires an identity this "own" refers to), not only to subordinate operative and strategic choices to *given* goals within set parameters but to transcend the very goal decisions and relate back to those goals by either autonomously acknowledging or rejecting them. For fundamental reasons, artificial systems will most likely never be in that position—even if they are certainly going to be very convincing in staging that they were—because that would require no less than a post-biotic conscience. For instance, such a system could be able to authentically tell us: "I could take you there, I could do this for you, but I don't feel like doing it today; I am not in the mood, and I'd rather like you to try something new with me or leave me alone for a while."

To summarize: human actors can use tools, operate machines, and interact with and within intelligent systems; therefore, they delegate tasks and decisions to the systems on an operative and strategic level, but they are always the only authority capable of normative decisions that cannot be delegated to machines. This is not to say that these normative choices on the human side were not restricted at all. Goal choices heavily depend on preference systems and worldviews expressed in feasibility judgments. Technology has a major effect on changing feasibilities. As a medium, it co-determines what is possible or impossible on a fundamental level, and some formations of this potential (whether actualized by humans intentionally or not, or by any other causation) might have constraining effects like path dependencies, coping actions, and so on (see also the chapters in the part *Controlling things that control us*). The following examples of (highly) automated systems capable of machine learning, such as everyday assistants or social robots, have to be seen in light of this threefold concept of autonomy—despite their rhetoric and marketing promises stating quite the opposite. Technological progress as driven by digitization, interconnectivity, miniaturization, and so on can be viewed as a shift from craftsman to conductor, from simple things such as tools to intelligent systems, from no autonomy delegation to comprehensive delegation of strategic autonomy, from learning humans to learning systems.

2. Intermezzo: Literally?

In the context of this book, the question whether the key terms *learning, caring, revealing, controlling, handling* are used literally or metaphorically is nontrivial. Wordings like "Learning from things that learn from us" and "Handling things that handle us" raise the question whether the first and the second instances of the respective verb are both meant literally or if the apparent symmetry is produced by the metaphorical power of language allowing to cover up differences on the signified level with identity on the signifier level. If signifier and signified would coincide in those symmetric slogans, then a maximum of insight could be transferred from one instance to the other; what we know about learning and caring among humans could then be transferred to learning and caring in human-machine relations. This understanding would then imply that things could care about people just as people care about people. This would imply again that things could choose who to care about or not (like we choose friends and partners), when to stop caring (which happens even between parents and children), and how to weigh self-care and care for others against each other. Could things literally *control* someone even though they are not capable of setting normative goals that guide the control? Would they not be better understood as one fragment among others that co-actualize the control? Could things *reveal* someone in a literal sense even though revealing includes the intention of making people aware of something or someone, which presupposes a preference-based choice of who to make aware and who to leave in ignorance as well as a notion of secrecy, openness and hiddenness, exposure and concealment? Could things literally *handle* someone or something even though most things do not have *hands?* Is it not the case that a human *handler* becomes metaphorical when he or she handles people instead of things or things that are not *handy* or *ready-to-hand* (Ger. Zuhandenheit)? These questions seem to lead to the conclusion that the symmetry is indeed predominantly on the signifier level and that there are significant differences on the signified level. Nevertheless, the metaphorically produced proximity of human learning and machine learning, of human care and thing care, and so on reveals promising human-thing similarities. This chapter focuses on the concepts of handling and learning on both the human and technical side. So far, we have examined *handling things* (1) and *being handled by things* (2). What about *learning*?

3. Machine learning, everyday pervasion, and the cunning of systems

In technical acting, the first shift from direct control to comprehensive delegations is accompanied and partly enabled by the ability to *directly* learn on the technical side. However, is technology actually *learning* in a similar way to humans or is this a metaphorical use of the term?

Learning can generally be defined as the process of committing knowledge to memory, of acquiring and training skills, habits, and attitudes. It is a structural change that results in the creation of or change in behavior or action as a response to a given situation with relative durability. Relative durability is what makes the difference between learning and, say, a behavioral change due to fatigue or drug use, since these changes usually pass after a nap or when the drugs wear off. This definition seems to be applicable to humans as well as to systems.[7] Humans include information in existing knowledge structures and commit newly acquired knowledge to memory, whereas technological devices save information on a hard drive locally or in the cloud. Of course, these are actually completely different processes: people do not save their impressions in a file system and artificial systems do not have any knowledge at all, but rather data, or information at best. But the notion of a structural change—be it in neurons or code—with relative durability, resulting in new or changed behavior, actions, processing, or output, generally seems to justify a comparison between both machine learning and human learning despite all differences. In technical acting, knowledge or information is essentially about the suitability of strategies, methods, and resources, as well as the skills of actualizing, executing, and applying them.

The change in machine learning, which can be seen in direct structural changes on the technical side, is a major driver for the above-mentioned shift in human-technology relation types and is at the very core of highly automated, "autonomous" intelligent systems. This learning process heavily depends on human-technology interaction exposure just as human learning depends on trial, error, and practice. Undoubtedly, test environments with training data enable systems to learn initially, but the massive progress in machine learning takes place when learning systems encounter millions of users in millions of situations and interaction scenarios. Therefore, the fact that intelligent systems pervade everyday life improves their learning conditions and accelerates their development.

Currently, one type of everyday life system with huge interaction exposure is smart speakers with all task assistance systems like Amazon Echo with Alexa, or Google Home. The number of units sold worldwide exceeded 40 million in 2016 and 2017. With an estimate of more than 50 million devices sold in 2018 alone, smart speaker home assistance systems are the fastest growing consumer-tech field (canalys 2018). Assistance systems on smartphones such as Apple Siri or Windows Cortana are a popular part of technologically assisted everyday life already, given that (as of March 2018) there are 8.5 billion mobile connections, over 5 billion mobile phone users, and among these mobile phones 50 percent are smartphones already—percentage rising (Statista August 2015; GSMA Intelligence 2018). Another type of learning system with everyday life assistance services are *social robots*, which are gaining momentum but are still relatively new compared to smartphone or smart speaker assistants today.[8] Those systems' signature feature is that they are capable of learning as shown by the following statements:

> Alexa—the brain behind Echo—is built in the cloud, so it is always getting smarter. The more you use Echo, the more it adapts to your speech patterns, vocabulary, and personal preferences. (Amazon October 2016)
>
> Meet JIBO, the World's First Social Robot for the Home.... So while he'll gladly snap a photo, he'll also get to know you and the people you care about. Every experience teaches Jibo something new, like recognizing the faces and voices of close family and friends, playing games, telling jokes, or sharing fun and interesting facts. And he's always learning more.... Jibo works hard to get to know his new family. And as he does, he becomes more and more a part of the funny stories, tender moments, and warm memories families share. Jibo is smart. And he's getting smarter all the time thanks to the incredible community of developers creating new, diverse skills for him to learn. (Jibo 2018)

> Pepper loves to interact with you, Pepper wants to learn more about your tastes, your habits and quite simply who you are…. Pepper adapts himself to you! Your robot evolves with you. Pepper gradually memorises your personality traits, your preferences, and adapts himself to your tastes and habits. (Aldebaran Robots 2016)

This learned adaption is one major development toward intelligent systems, as they not only learn how to differentiate a cup from a glass or one user from another but get to know the person they are serving (as represented by data). This is also called "personalization," which is a change in the system's structure and behavior due to accumulative user profiling.

By learning something about the human actor, the systems not only aggregate data in a data shadow and adapt their future processes accordingly but reveal the users via their data to a (paying) third party. This revealing often is the actual but hidden function of a system that offers apparent services to users that really are means to the revealing function. These systems—or the business models they are embedded in—only *simulate* to mainly serve the users as customers while *dissimulating* that the users—along with the simulated services—actually serve as goods being sold to other customers in the economic logics of a "surveillance capitalism."[9] Strategic part of these logics is the dynamic of simulation and dissimulation, of revealing and concealing, which can be seen as "a coup from above,"[10] or as a *ruse*. What Hegel describes as the *cunning of reason* ("List der Vernunft"[11]) is instructive to explicate this: An end relates to an object and, thereby, turns it into a means (to this end), which Hegel thinks of as an act of force.[12] The cunning of reason here consists in the fact that the end, relating to an object as a means, inserts another object between itself and the first object. This second object is a means to the end, but it appears as an end in relation to the first object.[13] This means that the end in means-end relations can function as means to a higher-level end, and so on. A ruse, more generally put, means an act that pursues another goal in addition to the apparent one. This other goal can be reached via the apparent goal (which is thereby a means to the other goal) but stays concealed—and this concealment is essential to its success.

Accordingly, the cunning of learning systems can now be seen as follows: A system offers a service (e.g. music streaming) and the actor chooses the system as a means to his or her goal (e.g. listening to music). By using the system's service as a means, the user not only reaches his or her goal but the user is converted into a means to a quite different end without being aware of it. This other end in the case of Spotify[14] is to gather and sell data on nothing less than the users' "moods, mindsets, tastes and behaviours" (Spotify 2018). "The more they stream [means to goal 1, BG], the more we learn [goal 2, BG]. This user engagement fuels our streaming intelligence [means to goal 2, BG]—insights that reflect the real people behind the devices" (Spotify 2018). The human actor is thereby converted from being an end into a means—according to Hegel an act of force. The cunning of systems[15] conceals this forced conversion—the user becomes useful and the buyer becomes the product. This concealment is a prerequisite to the success of this cunning function of revealing most private aspects of people's lives while concealing this process.[16]

Learning systems such as Alexa, Pepper, or Jibo provide a vast personalization of the *technosphere* (Ihde 1979). It would be a misunderstanding to think of a smart speaker such as Echo as just a device owned by the user. Echo devices are mere interfaces to the Alexa service that is connected to all kinds of sensors, databases, actuators, networks, and so on. Echo is actually the interface between the human user and the offered service. This human-technology relation is staged as an interaction similar to a human assistant hiding the vast background relations and interconnectedness that enables this interaction in the first place. This misunderstanding, which is intentionally provoked by the design (because it serves the cunning of systems), has severe consequences for the human actor and his or her future possibilities, as illustrated below.

At this point, it is important to bear in mind that the actual state of the art of intelligent systems is not yet where the marketing rhetoric of Amazon or SoftBank Robotics and so on claims and sells it to be. Yet, the combination of two tendencies indicates a development toward quasi-omnipresent, everyday personal assistance provided by learning systems: first, the fact that intelligent systems are currently pervading everyday life, especially with a tendency to multi- or all-task assistance with increasing delegation even on a strategic level—the plethora of smartphone apps or Alexa's skills being an indicator of the development of learning systems replacing one-task devices such as navigation systems;[17] second, an ongoing personalization—which actually creates increasingly specific stereotypes—of the system's behavior toward the user (user data shadow). Pepper and Jibo are about to grow up from being the toy-like learning entertainers they are today to capable all-task assistants, and they are likely to pervade every part of life from work and home down to social relations and sexuality.

One example might represent a link between the intelligent systems of today and tomorrow. It is the social domestic assistance robot Romeo, which is still in its prototype stage.

> After the meal, Romeo knows that Mr. Smith usually has a half-hour nap. Indeed, Mr. Smith goes to his bedroom. However, after one hour, he has not come out again. Romeo becomes concerned and enters the bedroom to check that all is well. He tries to speak to Mr. Smith but he does not respond. Romeo then contacts the remote assistance centre who take control of Romeo to assess the situation. As speaking is not enough to wake Mr. Smith, the remote operator takes control of Romeo's hand to shake him gently, while taking care not to injure him. This time, Mr. Smith wakes up. He was simply more tired than usual. After the children have left, Mr. Smith stays sitting in an idle position for a long time. Romeo becomes concerned and suggests activities based on

Mr. Smith's habits: calling a friend to play a game of cards, giving him a book or the TV guide. Mr. Smith opts for the game of cards and calls his friend. (Projet ROMEO 2017)

Again, the all-task assistance and the personalization are obvious here. However, is living with Romeo going to make us the Juliets of a future human-robot assistance society? Consciously or not, the Romeo Project partners (among them SoftBank Robotics) chose a scenario that bears resemblance to Shakespeare's fifth act, only that Robot-Romeo is trying to wake Mr. Smith (and not sleeping Juliet) and, instead of taking poison when not succeeding, he/it is dying the temporary robot death of tele-operated metempsychosis. If names and concepts matter and if art has an instructive power, then we might be wary of not reaching for the dagger when we find our Romeos less alive than expected. Let us not forget that Shakespeare's message was quite different from Mr. Smith playing cards happily ever after: "For never was a story of more woe/Than this of Juliet and her Romeo." The lesson here is that the role we assign to our artificial companions might change our role in turn, just as conversing with Echo—be it Amazon, the internet, or the technosphere as echo chambers in a broader sense—enforces the Narcissus in us.[18]

Effects on Future Openness

What are the effects on our future's openness (i.e. 'how open the future is', not 'frankness') when designing, creating, and interacting with learning things that constantly learn how to handle us? This chapter's main hypothesis is: Things get to know who we are and tie us down to who we were. To be more precise at this point of the argumentation: Things—intelligent systems such as Alexa, Jibo, Pepper, Romeo, and their successors—actually get to know who (they think or compile) we are and (they might) tie us down to who (they think) we were. A second look at Romeo's vision provides an exemplary starting point to underpin this hypothesis. "Romeo knows that Mr. Smith usually has a half-hour nap … " and—we might continue by combining the two already mentioned tendencies—tells Alexa to dim the lights, block incoming calls, close the garage, turn off all media devices, ignore the doorbell, activate the noise canceling systems (Quiet Comfort), heat the sheets, and what else smart homes and technosphere-connected personalized systems might do in order to proactively arrange the world according to our profile-based demand predictions. In other words: Things, both emerging and already-established comprehensive intelligent systems, gradually turn our world into a "new uterus"—warm and soft, compliant, in accordance with our preferences, providing, shielding, and never irritating. Is this the brave new world, a cozy technological womb of Eden?

So what is the problem with living in a technosphere of milk and honey? Some problems such as uncertainty, power imbalance, unequal risk-chance or cost-benefit distributions, injustice, unfairness, and nonsustainability have already been established and discussed. Someone has to actually cook and bring the pizza that

Alexa orders. This chapter focuses on a specific pitfall of personalized proactive techno-comfort that ultimately stuns our normative—the only nondelegable—autonomy.

There is and always will be an irreducible gap between the actual individual (Ger. *das reale Subjekt*) and his or her digital profile or data shadow (Ger. *das virtuelle Subjekt*). This data shadow is mainly but not exclusively a compilation of past interactions and past behaviors the system has learned from. But previous behavior, preferences, delegations, parameter settings, and so on are not to be perpetuated per se. The fact that someone did something does not mean that he or she wants to do it again. Systems learn from a digitized behavioral trail; while over-accentuating actual actions and neglecting intentional aspects, they completely miss concepts such as forced and reluctantly executed actions or the difference between normatively chosen and merely accepted heteronomous goals. Due to the expansion of the technical manipulation sphere that substantially transcends PCs or mobile devices (IoT, CPS, I4.0, etc.), the "world"—as in the technologically manipulatable sphere, such as info-sphere, service-sphere, interaction-sphere, and so on—is comprehensively pre-arranged. The potential formations of the technical medium are increasingly pre-formed or pre-actualized. Such world pre-formations are based on the compiled user profile or data shadow. Yet, this profile is not only derived from previous behavior of the real individual; the system also learns from every other user of the same *learning ensemble*[19] (e.g., all Echo users). Everything every single member of the learning ensemble teaches the system—or the systems reveal about the member—changes the world of every single user: Echo is not a device but an interface! "Alexa is always getting smarter… And because Echo is always connected, updates are delivered automatically" (Amazon 2018). This means that the personalization of a system's services is actually far from being focused on a single person, although this is exactly how the system's learning is promoted.

Personalization is really just advanced stereotyping—mostly.[20] This is not to say that the system does not get to know the individual (always by means of digitized traces of course). It means that the world's pre-formations and offered services are derived from more than one person. Furthermore, the system specifically learns from a user's "digital twins" or close "profile-relatives," meaning the group of all users that share a sufficiently similar data shadow (e.g. Caucasian, male, middle-age, unemployed, living in *x*, consuming *y*). What our digital twins do and do not do changes our world as manipulated by all systems that compile and access these twin profiles. In addition to that, the systems—although capable of personalization and learning—pre-arrange the user's world according to factors that are external to the systems and the compiled profiles, quite simply because the systems' operators, the Zuckerbergs, Pages, Brins, Pichais & Co., want to. They can push the featured information, services, and interaction options that they like most or that they are paid for. The options they (the operators and paying customers) prefer to change our world.

When confronted with a system's option, the individual user is unable to recognize which of these four factors influenced the offered option exclusively, predominantly, and to what extent. It is not in the operators' interest to let them know, which is why they stage their systems' services as a personalized result of only one individual. The uncritical user responds to the offered options in good faith as if they were results of his or her very own behavior and preferences.

As already mentioned, human actors are the only acting instances to autonomously choose goals, while technical things (tools, machines, systems of all sorts) choose—if at all—the ways and means. In terms of learning, gaining, or losing skills and competencies, gap exposure and practice are crucial: *use it or lose it*. Choosing the adequate means for a certain strategy requires competence in suitability judgment, as choosing adequate strategies to certain goals requires efficiency or effectivity judgment. Choosing adequate goals then requires feasibility and normative judgment. These operative, strategic, and normative choices are a matter of skill themselves, and they can be learned, forgotten, trained, improved, or diminished. If it holds true that even intelligent systems cannot autonomously set goals, then human actors still have to do so, despite all operative and strategic delegations, so they keep using their normative autonomy and therefore do not lose it.

So, what is the problem? To put it briefly: *if you get all you want, you need to get what you want*. More precisely: If you receive or if you are being offered "all you want," you need to understand what it is that "you want" first. Otherwise, you have to accept that what you get is what you wanted, especially if it is offered to you as derived from your very own preferences.

Normative or moral autonomy, the freedom to choose our own goals, the wish to develop, and the freedom to choose the direction and the will to change (to be more, something else, or better) are a prerequisite for *leaving the uterus* in the first place and to not re-implant ourselves in a new one. Personalized, learned, profile-based info-, service-, interaction-spheres tend to perpetuate old preferences (past goal choices) and, thus, tie the individual to their older digital version, to their former digital twins. Personalized world pre-formation narrows the gap between a person and his or her ascribed type, between a real individual and his or her data shadow, not by perfecting the shadow according to the individual but by molding the individual into the shadow's form. As this metaphor suggests, this might result in technologically more compatible humans such as predictable and constantly consuming customers; but in the long run strong individuals, *conductors*, leading an autonomous life knowing what they want will eventually end up being a *shadow of their former selves*.

Because goal choices are irritable. The pre-formation of the connected world—following digitally assumed goals—precedes the goals that are actually formed or expressed. This pre-formation is executed without the users being aware of it (cunning of systems). Recommendations—technologically assisted goal choices—interrupt the goal formation process, thus possibly stunning the goal-setting competence. Goal autonomy is not at risk because systems could autonomously choose for you; they cannot, and human normative decisions are rather disguised as system processes (see aspect 4 above).[21] It rather is at risk because the human actor's goal formation can be disturbed by proactive choice assistance and diverted into the direction of pre-formed surrogate goals.

Magic wish fulfillment as described in the land of milk and honey and as promised by today's comfort systems[22] tends to backfire. Hannah Arendt points this out in the context of eliminating labor by automation:

> "The fulfilment of the wish, therefore, like the fulfilment of wishes in fairy tales, comes at a moment when it can only be self-defeating… society does no longer know of those other higher and more meaningful activities for the sake of which this freedom would deserve to be won" (Arendt 1998, 4–5).

What is the use of a wish when all you can imagine as desirable are more wishes? Ortega y Gasset predicts a crisis of wishing for the technologically empowered man and links technological power to missing normative capabilities. This crisis of wishing can be viewed as a crisis of normative autonomy, as a crisis of goal choice competence[23]:

> He (man the technician; *el técnico*) holds in his hands the means to achieve his wishes, but he does not know how to wish. At heart he notices that he wishes nothing, that he is incapable to orient his desire and to decide among the countless things that the environment offers to him on his own. Therefore, he looks for an intermediary that orients him and finds it in the predominant wishes of others. That is the reason why the nouveau riche first buys a car, a player piano, and a phonograph. He puts others in charge who wish for him. (my translation of: Ortega y Gasset 1964: 343–344)[24]

If he had known today's highly automated systems, Ortega y Gasset could have painted the picture of a *conductor* of the world's best orchestra who is absolutely clueless about how he or she wants it to sound. It is worth examining this quote more closely, as it connects several aspects of this argumentation. *El técnico*, man, the technician, a human actor with technology at his or her disposal, developed from craftsman to conductor and in current human-technology relations he or she mainly appears in the role of a *user*. The fact that he or she "does not know how to wish" indicates that the user has no E_i, no internal end, no imagination of what meaningful activity to realize with his or her technological power. The sought-after "intermediary that orients him" or her comes in the form of recommendations (e.g. Spotify's, YouTube's and Netflix's recommendation systems) and in the form of unsolicited pre-formations such as Romeos "time for your nap Mr. Smith." The sought-after intermediary is found "in the predominant wishes of others." Given the date of Ortega y Gasset's quote (originally written in 1933), he could not have known about today's highly automated systems, but today, the predominant wishes of others are exactly what trains the system and what changes how the system pre-forms the world. These *others* are the older selves, the digital twins, and the connected learning ensemble (e.g. Amazon's "Customers who bought this item also bought" section or every Top Ten, Most viewed/listened/bought list).

The user has the means to accomplish almost every objective (i.e. almost *all* E_e), but he or she lacks normative orientation or goal choice skills (i.e., *no* E_i), so he or she "puts in charge the others who wish for him" and finds these others staged as his or her own personalized options. The point transcending Ortega y Gasset in the context of today's intelligent systems is that the normative heteronomy, or goal cluelessness (no E_i), does not just coincide with unprecedented technical realization might (almost all E_e); rather, the latter is causal to the former. The formation of goal choices E_i is hindered by the interruption of offered Ee (recommendations) and by the proactive pre-formation of the world (situation, connections, filters, context, likes, appreciation etc.) in which the E_i could form. The pre-formed world is narrowed down in terms of future openness, allowing less development in contrast to previous choices and outside their gravitational field. The freedom *of* choice on the top normative level depends on the freedom *from* predetermining or manipulating elements. As a consequence, genuine E_i—not systemically proposed, or artificially assisted goal choices—vanish and, thus, the instructive difference between E_i and E_e. Being aware of this difference is a crucial condition for the possibility to learn, as it provides information on suitability and feasibility choices, execution qualities, and so on. This difference reveals not only the suitability of means, the feasibility and desirability of goals but also aspects of the determining mediality that, as a structured possibility space, grants and refuses options of means and goal identification and choice in the first place.

Proactive, personalized choice assistance can stun the normative autonomy and change it to a normative heteronomy where the human actor unknowingly follows "the wishes of others." Thus, the conditions for (a) learning how to handle things—which is not always a problem—and (b) learning how to "autonomously" form "wishes" or develop goals worth striving for become worse. Eventually, in a long-term perspective, the difference between internal and external ends or goals that is pivotal for learning and development on the human side, could turn into the nondifference of external and internal goals—meaning that, by being realized, the external goals dictate what the internal goal had to be in the first place. What you get is what you wanted; therefore, you always get what you want, and therefore you never fail in getting what you want: realized and intended end ultimately coincide and, thus, withdraw both the possibility and the need to improve. A condition in which neither improvement nor development nor the potential to be more, something else, better, or different is neither necessary nor possible is truly a condition with minimal future openness or no future at all.

In this context of current and emerging, pervasive, proactive, and personalized intelligent systems and in terms of effects on human actors' normative autonomy, future development, and human-technology-world relations, we can summarize: *the more things learn, the less we learn* (regarding the same task). The more things know about us and the more they proactively pre-arrange our world, the fewer options we have to differ from our past digital selves. In this sense, *handling things that handle us* means *things get to know who we are and tie us down to who we were.*

Notes

1. This also involves knowing the difference between a feasible end and a mere wish. Means and ends are codependent: being realizable by means—feasibility— determines an end (an unrealizable end would be a mere wish), and vice versa, leading to an end—suitability—determines a means. See Hubig (2006: chapter 4).
2. Of course actors lose skills they once possessed just as they gain new ones. Acting is not at all an all-improving development. Skills are built and lost because of complex factors, one of which will most certainly and crucially be practice. As the saying goes: *Use it or lose it*. In the perspective proposed here, "practice" is nothing else than gap exposure.
3. See, for instance, the acatech levels of automation in the field of automated road traffic (acatech 2015).
4. As the normative autonomy is not delegable to technology, only human actors can choose to acknowledge a given or to freely choose any other goal. Artificial agents are restricted to given goals.
5. Some might reject this "in principle" judgment, but a moral autonomy of the here-mentioned kind on the system's side would require no less than an artificial post-biotic consciousness (fully realized and not just simulated). The "in principle" judgment changes depending on how feasible or actually possible such a consciousness is deemed. As for today at least, there is no doubt that such a consciousness is not possible, and therefore it is impossible to delegate moral decisions to machines.
6. This was already proposed by Gilbert Simondon in 1958: "Far from being the supervisor of a squad of slaves, man is the permanent organizer of a society of technical objects which need him as much as musicians in an orchestra need a conductor. The conductor can direct his musicians only because, like them, and with a similar intensity, he can interpret the piece of music performed; he determines the tempo of their performance, but as he does so his interpretative decisions are affected by the actual performance of the musicians; in fact, it is through him that the members of the orchestra affect each other's interpretation; for each of them he is the real, inspiring form of the group's existence as group; he is the central focus of interpretation of all of them in relation to each other. This is how man functions as permanent inventor and coordinator of the machines around him. He is among the machines that work with him." (Simondon 1980, 4).
7. Whether this application to humans and systems is justified or not depends on how *learning* is conceptualized. Is it defined as something that humans do when they durably change their behavior in response to certain conditions or events? Then the application to artificial systems is a metaphorical one, which still offers valuable insights into "learning" systems. However, this non-literariness has to be kept in mind in order to not miss relevant differences between both. Is learning defined as a durable change in structure and behavior as a reaction to external stimuli? In that

8   case, it is a more abstract concept in which human learning and machine learning are two subtypes just as animal, plant, (un-)animated nature learning, etc.
8   According to SoftBank Robotics Corp., who sell the Pepper Robots, they sold 1,000 units within one minute in 2015 SoftBank Robotics Corp (2015).
9   Shoshana Zuboff defines *surveillance capitalism* as: "1. A new economic order that claims human experience as free raw material for hidden commercial practices of extraction, prediction, and sales; 2. A parasitic economic logic in which the production of goods and services is subordinated to a new global architecture of behavioral modification; … 8. An expropriation of critical human rights that is best understood as a coup from above: an overthrow of the people's sovereignty" (Shoshana Zuboff 2019).
10 Ibid.
11 *Ruse* and *cunning* is meant as in the German "List," just as Hegel used it in the infamous phrase "List der Vernunft" (G.W.F. Hegel: Die Wissenschaft der Logik: Part Two—Die subjektive Logik, Section Two—Die Objektivität, Chapter Three—Teleologie).
12 Ibid.
13 Ibid.
14 Special thanks to Heather Wiltse for pointing me to this "for brands" aspect of the Spotify example.
15 This *cunning of systems* itself is a metaphor in this case because contrary to the cunning of reason, the systems themselves do not set up this cunning ruse, but the people behind the systems do, the owners and operators in charge.
16 For further thoughts on the role of *revealing*, see also the section *Revealing Things That Reveal Us*.
17 You obviously cannot delegate your music choices or scheduling tasks to a navigation system, but you can navigate it with a smartphone and you can use Alexa as a secretary or for ordering "favorite food," etc.
18 See the comparison between the mythical Echo and Amazon's Echo in "Assisting ourselves to death. Unruly artificial assistants as competence sensitive enablers" (Gransche 2018).
19 Another example of such learning ensembles is the group of Baxter robots in the *Million Object Challenge*: The research team of the *Human to Robots Laboratory* of Brown University attempts to teach Baxter robots how to manipulate 1 million things in *The Million Object Challenge* (Brown University 2016b). A multitude of robots experimentally learn how to grab objects and store the acquired skills in a database. By the end of the experiment, every robot connected to this pool has learned the skills to manipulate 1 million objects, even if one component has only learned a few or none of these skills by experimentation. It is not really surprising that the *Humans To Robots Laboratory* team uses Amazon Echo as a voice control interface to interact with Baxter robots as well (Brown University 2016a). See also Gransche (2018). Alexa & Co.'s actions resemble a *Billion Person Challenge*.
20 This holds true not so much for the data gathering that can actually be derived from an individual user's traces (even though the framing, selection, filtering, aggregation, etc., of the data is super-individual again). This holds true especially for the abductions that link the learned data to strategic decisions: What does the fact that a person is streaming Breaking Bad, The Sopranos, and Batman—The Dark Knight Rises (for the eleventh time) mean for the chances he or she will buy my smart toaster or dental insurance?
21 "Digital information is really just people in disguise" (Lanier 2013: 15).

22   Here, the promise is that you can act without any effort, without lifting a finger, and Amazon and Google suggest a fairy-tale-like magical easiness, only that the magic word is not *Open Sesame* but *Ok Google* or *Alexa*: "When you want to use Echo, just say the wake word 'Alexa' and Echo responds instantly" (Amazon October 2016). "Use Echo to switch on the lamp before getting out of bed, turn on the fan or space heater while reading in your favourite chair, or dim the lights from the couch to watch a movie—all without lifting a finger" (Amazon October 2016).

23   A distinction between an end or goal on the one hand and mere wishes on the other hand was proposed above, namely in the aspect of *feasibility*. A goal that is (correctly) judged as unfeasible is not a goal but a mere wish. Ortega y Gasset clearly uses *wishes* in the feasible sense here, because here the technician "holds in his hands the means to achieve his wishes." An achievable wish is a goal or an end. Therefore, Ortega y Gasset's crisis of wishing can be read as a crisis of goal choice competence.

24   Original quote: "Tiene en la mano la posibilidad de obtener el logro de sus deseos, pero se encuentra con que no sabe tener deseos. En su secreto fondo advierte que no desea nada, que por sí mismo es incapaz de orientar su apetito y decidirlo entre las innumerables cosas que el contorno le ofrece. Por eso busca un intermediario que le oriente, y lo halla en los deseos predominantes de los demás. He aquí la razón por la cual lo primero que el nuevo rico se compra es un automóvil, una pianola y un fonógrafo. Ha encargado a los demás que deseen por él" (Ortega y Gasset 1964: 343–344).

References

acatech. 2015. *New autoMobility: The Future World of Automated Road Traffic*. acatech Position.

Aldebaran Robots. 2016. *Who Is Pepper?*. https://www.aldebaran.com/en/cool-robots/pepper (Retrieved March 10, 2016).

Amazon. 2016. *Amazon Echo*. https://www.amazon.com/Amazon-Echo-Bluetooth-Speaker-with-WiFi-Alexa/dp/B00X4WHP5E (Retrieved October 30, 2016).

Amazon. 2018. *Echo Spot: Alexa-Enabled Speaker with 2.5" Screen — Black*. https://www.amazon.com/Amazon-VN94DQ-Echo-Spot-Black/dp/B073SQYXTW (Retrieved March 8, 2018).

Arendt, H. 1998. *The Human Condition*. 2nd ed. Chicago: University of Chicago Press.

Brown University. 2016a. *Amazon Echo + ROS + Baxter*. http://h2r.cs.brown.edu/amazon-echo-ros-baxter/ (Retrieved November 1, 2016).

Brown University. 2016b. *Million Object Challenge*. http://h2r.cs.brown.edu/million-object-challenge/ (Retrieved November 1, 2016).

canalys. 2018. *Smart Speakers Are the Fastest-Growing Consumer Tech; Shipments to Surpass 50 Million in 2018*. https://www.canalys.com/newsroom/smart-speakers-are-fastest-growing-consumer-tech-shipments-surpass-50-million-2018 (Retrieved March 5, 2018).

Gransche, Bruno. 2018. "Assisting Ourselves to Death: Unruly Artificial Assistants as Competence Sensitive Enablers." In *Philosophy of Engineering: Foundations of Futures*, edited by Albrecht Fritzsche and Sascha Oks, 271–289. Berlin: Springer International Publishing.

Gransche, Bruno, Erduana Shala, Christoph Hubig, Suzana Alpsancar, Sebastian Harrach 2014. *Wandel von Autonomie und Kontrolle durch neue Mensch-Technik-Interaktionen: Grundsatzfragen autonomieorientierter Mensch-Technik-Verhältnisse*. Stuttgart: Fraunhofer Verlag.

GSMA Intelligence. 2018. *GSMA Real-Time Tracker Mobile Devices*. https://www.gsmaintelligence.com/ (Retrieved March 5, 2018).

Hubig, Christoph. 2006. *Die Kunst des Möglichen I. Grundlinien einer dialektischen Philosophie der Technik; Technikphilosophie als Reflexion der Medialität*. Bielefeld: Transcript.

Ihde, Don. 1979. *Technics and Praxis*. Dordrecht: Reidel.

Jibo. 2018. *Jibo—The World's First Social Robot for the Home*. https://www.welcome.ai/products/robotics/jibo (Retrieved March 5, 2018).

Lanier, Jaron. 2013. *Who Owns the Future?*. London: Allen Lane.

Ortega Y Gasset, José. 1964. *Obras Completas: Tomo V (1933–1941)*. 6th ed. Madrid.

Projet ROMEO. 2017. *Projet ROMEO*. http://projetromeo.com/en (Retrieved June 13, 2017).

Schilling, M., S. Kopp, S. Wachsmuth, B. Wrede, H. Ritter, T. Brox, B. Nebel, and W. Burgard. 2016. "Towards a Multidimensional Perspective on Shared Autonomy: Technical Report FS-16-05." *The 2016 AAAI Fall Symposium Series: Shared Autonomy in Research and Practice*.

Simondon, Gilbert. 1980. *On the Mode of Existence of Technical Objects*. Paris: Aubier, Editions Montaigne, 1958: University of Western Ontario.

SoftBank Robotics Corp. 2015. *1,000 "Pepper" Units Sell Out in a Minute*. https://www.softbankrobotics.com/emea/zh/1000-pepper-units-for-july-sell-out-in-a-minute. (Retrieved March 5, 2018).

Spotify. 2018. *Spotify for Brands. Audiences*. https://spotifyforbrands.com/en-GB/audiences/ (Retrieved March 5, 2018).

Statista. 2015. *Number of Mobile Phone Users Worldwide 2013–2019*. https://www.statista.com/statistics/274774/forecast-of-mobile-phone-users-worldwide/ (Retrieved March 5, 2018).

Zuboff, Shoshana. 2019. *The Age of Surveillance Capitalism. The Fight for a Human Future at the New Frontier of Power*. London: Profile Books.

5 CAN ETHICS BE LEARNED? VIDEO GAMES AS AN ETHICAL SANDBOX

Fanny Verrax

Learning Ethics: The Self and Others

Whether ethics can be taught and learned is a sensitive and thoroughly discussed question. Most of the literature usually focuses on academic or vocational training in ethics to assess its practical benefits (Geary and Sims 1994; Ritter 2006; Langlois and Lapointe 2010). This chapter chooses to take a slightly different perspective: the amount of ethical learning enabled by video games, and more specifically, single-player video games—that is, video games in which there are no direct interactions between players. Hence while aiming at answering the question "can ethics be learned?," which is central in moral philosophy, the chapter will also address a question more specific to philosophers of technology: How necessary are human interactions in order to facilitate the learning of a typical human skill, and what is the role of things that relate to us in this learning process?

The discussion is still vivid in ethics on what should be the very object of ethics: one's relationships to other humans only—the position known as minimal ethics—or one's relationship to the self also—the position known as maximalist (Ogien 2007), while non-anthropocentric views emphasize duties to nonhuman others. This chapter does not intend to enter this debate but rather to suggest the idea that nonvirtual others are not necessarily needed in order to foster an ethical learning, which can also benefit relationships with humans.

When video games are praised in game studies and related fields, it is usually for one of two reasons. First, their ability to promote development of cognitive skills, such as memory and attention, is acknowledged—sometimes to the point that it is said to outgrow that of reading. Steven Johnson, notably relying on the work by James Paul Gee, suggests for instance that the typical activity involved in most video games (probing, forming a hypothesis, reprobing, rethinking) is very similar to the scientific method and can be seen as a great way to learn it (Johnson

2006). Second, when it comes to analyzing interactions, the focus is on the communication with nonvirtual others, through the video game as a medium: "a fundamental part of the process of developing our moral understanding of games is belonging to a game community, experiencing the presence of and interacting with other ethical beings who play computer games" (Sicart 2011, 9). This kind of analysis is particularly developed for MMORPG (massively multiplayer online role-playing games), whose most famous examples are games such as *World of Warcraft* or *Dofus*.

But this is not our focus. I would like to argue here that single-player video games (video games in which the player interacts solely with an artificial intelligence) offer a unique experience of virtual alterity and ethical learning. This experience is of a different kind entirely, but as, if not more, interesting in terms of ethical learning than MMORPG and other video games relying on the multiplicity of nonvirtual players. More specifically, I would like to focus on two features of single-player video games that make them a worthy ethical experience: they expand moral imagination, and they provide a safe learning environment.

Characterizing Video Games

Games and Video Games

Before being characterized by their digital environment, video games are games. Dutch historian and anthropologist Johan Huizing identifies five fundamental features of games that are, consequently, also present in video games:

1) Freedom: Games are free in the sense that at any time, a game can be delayed or canceled. The need for playing exists insofar as there is a need for pleasure, but it never stems from a physical urge or a moral duty.

2) Culture: Games have a cultural function; they create spiritual and social bonds.

3) Isolation: Games exist in a separate time and space framework. This isolation allows for another essential feature: repetition.

4) Order: Games are structured by rules. "As soon as rules are violated, the game universe collapses" (Huizinga 29).

5) Tension: This means not only that there can be an element of uncertainty or luck but also that "something must succeed at the cost of a certain effort" (Huizinga 28).

Video games, just as traditional games, are therefore characterized by these five features: freedom, culture, isolation, order, and tension. Let me now be more specific about what makes video games specific, compared to non-video games.

NB. Given the extreme diversity of both video games and IRL games, these differences are to be understood not as radical differences but rather as a continuum.

How to Characterize Video Games

Premise n.1: Video Games Are Ergodic Activities
The seminal work of one of the founders of humanistic informatics, Norwegian Espen Aarseth, allows us to think of video games as an *ergodic* (from *ergon*, work, and *hodos*, path) activity, in the sense that the sequence of signs changes according to the player's decisions. This is what Aarseth calls a "non-trivial reading" that is core to cybertext and video games (Aarseth 1997). I argue this is a specific feature of video games in the sense that in traditional games, such as chess or checkers, and board games, such as *Risk*, *Monopoly*, *Settlers*, *Alhambra*, obviously players' decisions impact the economy of the game and how it looks, but the global sequence of signs remains the same. Even in more evolved non-video games, such as *Dominions*, it is true that the sequence of signs (here cards) changes from one time to another, but this happens *before* the actual game starts. This ergodic feature reaches its paroxysm in procedurally generated video games, such as *Minecraft* or many survival games.

Premise n.2: Video Games Foster Emotional Involvement
Beyond the *work* that is necessary to make sense in video games, it is the emotional involvement that characterizes computer games: "It makes a difference if we have to arrange blocks in an optimal position or if we have to save the princess from the jaws of a monkey" (Pohl 2008, 101).

Premise n.3: Video Games Allow for Moral Emotions
The issue of moral emotions is a hot one in ethics. It is not the point here to enter the debate on ethical intuitionism (Roeser 2010; Steinbock 2014) but rather to establish as a starting point for discussion that:

a) Moral emotions (or higher-order emotions), such as shame, guilt, embarrassment, and pride, exist and are experienced regularly in ordinary life.

b) Moral emotions can arise in video games as well, thanks to their fictional content: "without fictional context, the player actions cannot be interpreted morally" (Švelch 2010, 56).

Premise n.4: Avatars Are Not Indispensable to Emotional and Moral Involvement
Švelch (2010) focuses on avatar-based single-player video games for he believes the avatar is the seat of the player's agency, thus fostering moral engagement. Very simply put, the scheme that Svelch endorses is the following:

avatar → agency → moral engagement.

Although I do agree that avatars can provide a unique experience of virtual moral agency, I do not believe they are absolutely necessary for moral engagement.

Thus most of the examples in the chapter will be taken from non-avatar-based single-player video games, mostly simulation and strategy games, a realm underexplored by the literature.

The Ethics of Video Games: Experience Comes First

Little explored, the ethics of video games usually point to the problems they provoke rather than the solutions they can offer. In his chapter "Philosophy and ethics of game," Lafrance suggests for instance five ethical features that are all negative: futility, agonic character (competition), illusion, propagation of sexist stereotypes, violence (Lafrance 2006). Some games score particularly high on these scales—one can think of the delusional, highly violent, and sexist game *Grand Theft Auto* as a paradigmatic example.

But independently of what a game's particular content is about, one shall not forget that a game is never just a set of rules or a code, but foremost an experience, which is why it is important to distinguish between "games *in potentia*" and "games *in actio*" following the Aristotelian dichotomy.

> When reading the criticism some games like *Grand Theft Auto: San Andreas* has received for its violent content, media and game critics seem to focus only on the analysis of the ethical affordances of the game as a possibility. Ultimately, a game is not the object we describe when we write about the rules and the fictional universe, but the experience constructed by the interaction of a user with that world. In a sense, *Grand Theft Auto: San Andreas*, only exists as a moral experience when played, while it certainly is a moral object of incomplete nature when only described. Games from their design are moral objects, but we need to consider how they are experienced by players in order to fully understand the ethics of computer game. (Sicart 2005, 15)

In the video games that will be discussed throughout this chapter, we will try to see what is needed in order for the player to experience her own "game in potentia," which is a requirement to foster a proper ethical experience.

Video Games Expand Moral Imagination

Moral skills usually break down to two broad categories: judgment, including argumentation and deliberation, and imagination. Moral imagination is paramount to *perceiving* what moral issues are at stake in a given situation. This is why some philosophers refer to it as "imaginative moral perception." Canadian philosopher Martin Gibert identifies three modes of imaginative moral perceptions:

Perspective shift: The ability to change perspective, for example, by imagining somebody else's perspective on a given situation.

Reframing: Imagining a particular element of a situation as different from what it actually is.

Comparison: Imagining a whole situation as different from what it actually is (Gibert 2015).

I argue video games expand imaginative moral perceptions through these three modes. The next section will, however, particularly focus on the first mode, perspective shift. I will show that by providing the player with an experience of power, weakness, and overall diversity and multiple identities, video games allow for a wide range of perspective shifts.

Experiencing Power

> *The designers and fans of world-builder games are not exaggerating unreasonably or (at least in any simple sense) exhibiting impiety when they refer to world-builder games as "god games". The avatar's power over the world of the game is superhuman. And astute players and programmers of these types of games have wrestled with the moral quandaries that such power raises. How should someone with this kind of power rule? (Cogburn and Silcox 2008, 73)*

Citizens' lack of involvement in public life is often interpreted as a logical consequence of the very real lack of power to change anything for most citizens: the typical "why would I bother if I can't change anything?"

In the video games realm, world-builder games (*Civilization, Tropico*), city-management games (*Simcity, Cities sylines*), and political simulations games (*Democracy 3*) make it worth it to ask real-life political questions like the following:

—Should I create neighborhoods with different infrastructures and possibly regulations (such as curfew) based on income level? (*Cities Skylines*)

—Should I legalize gambling and give a permit to build a casino to generate extra cash, while knowing it will in all likelihood increase criminality in the neighborhood and may foster protests? (*SimCity*)

—Should I attack my neighbor before they attack me, or try to sign a pact of nonaggression? (*Civilization*)

—Should I raise the oil tax, knowing that commuters and car owners will be angry, but environmentalists will be happy, and that I have an asthma epidemic to deal with, plus I could really use the extra cash? (*Democracy 3*)

Issues that are addressed are not just present-based but span throughout history, up to sci-fi scenarios:

—Should I adopt slavery as a social policy, making it possible to finish buildings faster by sacrificing a part of my population? (*Civilization*)

—Should I try and live in harmony with aliens I do not quite understand yet, or destroy them preventively? (*Civilization Beyond Earth*)

Furthermore, the player can question the question itself: Why is it that building a casino in my city is going to increase the criminality rate? Is that even true? Why do I not have other options? This is where *mods* enter the game: from players who are not entirely satisfied with the game as it is and want to have more options. Mods are the way in which video games take their ergodic characteristic a step further: not only is the sequence of signs modified but the actual code is also modified.

Example: mods in *Democracy 3* include not only new countries but new social policies and new interactions between them. For instance, a mod adds "freedom of speech" and relates it to youth's satisfaction.

If some players are willing to spend hundreds of hours developing new mods to expand the horizons of their favorite video games, it is not entirely unlikely that this dedication could, someday, be transferred to nonvirtual issues and modes of actions (signing a petition, going to vote, changing their consumption habits, etc.) even though, obviously, empirical data would be needed to confirm this possibility.

Another way in which video games offer a unique decision-making experience is by presenting the player with specific professional contexts. Typical examples here include being a prison manager (*Prison Architect*), running a pharmaceuticals company (*Big Pharma*), or working as a customs officer in a totalitarian state (*Papers please!*). Here the moral dilemmas are quite explicit and clearly intended in the game design.

Some games also have the unique feature of offering the players both roles: the decision-maker and the one the decision is made for. In *Prison Architect*, for instance, the player can build and manage a prison as a warden and then enter "Escape Mode" and choose to be a prisoner. Certainly the focus of the game is on security: making sure as a warden the prisoners cannot escape, and as a prisoner finding the loophole that will allow for escape (if one plans it ahead as a warden, of course it is no fun). But there is something else to it: the player has to live the life of the prisoner, including the regime that has been decided for her: when she can eat, sleep, phone her family, and so on. Obviously I am not suggesting that this experience is similar to the one of actually being detained in jail, but I do believe it can still offer a valuable basis to escape a purely theoretical debate on these issues.

Experiencing Weakness

Indeed perhaps even more importantly than experiencing almighty power, some video games provide the player with an experience of **ethical weaknesses**, be it embedded in the game (*Papers please*) or be it derived from an experience of power abuse, when I, as a player, have the experience of being weak to the call of power (all world-builder games—called "God games" for a reason—can trigger that feeling to some extent).

I suggest there are two ways in which a human player can experience ethical weakness while playing video games. The first one has to do with failing to master a set of skills or failing to do it in time—this is mostly true in real-time games (as opposed to turn-based). For instance, in the Microsoft series' *Age of Empires*, a player can add for oneself a special ethical challenge and decide before starting a game that whatever happens, they will not kill the enemies' civilians (workers and priests). But as the game runs, the player can fail to achieve this goal, as the virtual army she commands has some autonomy, and will for instance attack a priest who is trying to convert them. This failure has to do with a lack of rapidity or dexterity ("I knew what I wanted to do but I didn't manage"), oblivion or negligence ("I forgot my army was there"), sometimes excess of pride ("I didn't think I needed to pause the game and I thought I could figure it out while running, but it turns out I couldn't").

Now, consider the constraints imposed upon the player of turn-based world-builder games, such as the impossibility of reacting immediately to the enemy's movements, or the constraints imposed by real-time interfaces, such as the difficulty of simultaneously managing the movements of hundreds of soldiers through the game-world. Might it be possible to obtain some insight into the nature of potential conflicts by reflecting upon how these limits forced on the player influence the coherence, realism, or even the entertainment value of god games?

If this example is based upon an extra challenge that perhaps few players actually deal with (although setting one's goals independently of the game's explicit missions is an essential feature of simulation games), some real-time games are designed to challenge the player's technical skills (memory, dexterity, visual recognition, etc.) while presenting her with explicit moral stakes. The game *Papers, please!* is an independent simulation game in which the player embodies a customs officer from a fictional soviet republic. While the set of rules to respect grows every day, the player also has to make calls: Shall she let pass this poor woman who seems harmless although she does not have correct identification? The woman may be a terrorist after all. And even if she's not, letting her through means the player has used up her mistake allowance quota. Next time, she may bring less money home and not be able to feed her family. The ethical aspect of the game therefore lies in its very gameplay, as put by one reviewer:

> But the moral quandaries are so real, so relatable that your conscience speaks to you more than it does in most other games. The decisions feel like they matter. Even if you're the kind of person to always take the "lawful good" route, to always take the high road in games, you're forced to commit acts for which you might hate yourself. If you don't, you and your family will not survive the winter. The fact that a few sentences of text accompanied by retro-style graphics can make you feel as bad as you do is impressive.
>
> (Britton Peele, Gamespot, August 13, 2013: https://www.gamespot.com/reviews/papers-please-review/1900-6412914/retrieved March 12, 2018)

In any case, the player can feel not only disappointment but also actual shame or guilt, including for mistakes that are purely technical, like not going fast enough, which is no different from real-life experience: if a driver hits a walker because they failed to turn the wheel fast enough, the fact that it was a technical shortcoming, and not a moral one, does not necessarily prevent the guilt.

The second way in which a player can experience ethical failure is by being weak to the call of power. This provides a particularly unique experience when the game is set in a historical context that the player knows at least a little about. It is the case for instance with the scenario "Scramble for Africa" included with the Brave New World extension pack of *Civilization V*. The player can choose to play a European power (France, Belgium, England, etc.) or an African country (Egypt, Morocco, Ethiopia, etc.), all starting in 1880. The victory conditions are different for European powers and African countries. If the player chooses to play a European power, she will be astonished at how easy it is to defraud commercially or conquer militarily the African territories, based on intercultural differences. Rights of way through another player's territory, for instance, are usually a quite expensive favor to obtain in the standard game. One must very often give significant amounts of resources or money in order to gain a right of way. But what is implied in *Scramble for Africa* is that African people do not hold the same conception of a "national territory" and therefore do not care, granting right of way without realizing its strategic value. The player is therefore facing the following question: Why offer a fair deal when the other player doesn't even realize she is being ripped off?

Another example is the so-called "dystopian business-simulator" *The Founder*. Created by Francis Tseng to mock the start-up world, the player is confronted with start-up founders' choices. One of them is choosing the kind of things she can say to potential employees in order to negotiate their salary as low as possible. Possibilities include sentences like: "We're building products that will change the world" or "You'll have a huge impact on people's lives here," although this does not affect the company's actual choices. Here the player is facing the question: If my employees agree to make less money as long as I sell them a harmless yet untrue narrative about their jobs, why shall I not enjoy it and make more profit?

Experiencing Multiple Identities

Finally, through all these various contexts, the player is invited to embody different identities. I argue that the type of moral engagement depends upon the kind of identity that is experienced.

In *Prison Architect*, for instance, one indicator that players can choose to attend to is the repeat offender rate. This rate is calculated from other indicators, such as prisoner's health, punishment intensity, and reform programs. Concretely, having the prisoner locked in her cell many hours a day and having armed guards patrolling often in front of the prisoner's cell will increase the punishment intensity, while offering general education classes and the possibility to work in the prison (laundry, kitchen, etc.) will increase the reform grade. So the dilemma is, for every player, whether to increase punishment at the cost of reform, or vice-versa, knowing that, according to the game's algorithm, the repeat offender rate will not be satisfactory if both are not taken into account. This is when the player's own values and beliefs often enter the game.

Švelch (2010) quotes two radically opposed views on the matter, that also emphasize the uttermost importance of the type of game considered:

> "When I play a BioWare role-playing game, my characters tend to not only lean toward the nicer side but almost immediately start twinkling with the magical pixie dust of purity. It's embarrassing, but I just make the decisions I believe I'd really make, and end up that way" (John Walker, Eurogamer. Com; Walker 2009).
>
> "I laugh out loud when I run pedestrians over in Grand Theft Auto and get a kick out of unleashing Godzilla on my Sim City. In fact, I can't name a video-game that did evoke any sadness or true ethical dilemma in me until BioShock" (Osama, TowardsMecca.com; Osama 2008).

In this perspective, I suggest the following typology of identities in video games:

Phenomenological identity: The appearance of the avatar the player embraces. The player can choose to make it resemblant to her actual looks, or not. Many games do not represent any avatar (such as *Democracy 3*: all the player can see is the back of a leader who is probably male); some have but one option per character choice (typically in *Civilization* the avatar depends on the chosen civilization and depicts an historical leader. Interestingly enough, the player hardly ever sees her own avatar, as its main function is to appear in the dialogue box when communicating with other civs); some offer a very limited set of choices that make it virtually impossible to choose a resembling avatar; some offer various choices, but none of which is based on human looks (in this category we will find mostly fantasy games or humorous games, such as *Girls and Robots*); and some finally offer millions of possibilities through a character design process that is a big part of the game (the most typical example is *The Sims* and all its extensions).

Ontological identity: The self-claimed personality characteristics of the avatar. The player usually must choose between a set of options that can be wide (*The Sims* is here again the most prominent illustration, with dozens of personality traits that can be combined. In the *Sims 3*, sixty-three personality traits can be combined to form hundreds of millions of possibilities. Examples of traits include "ambitious," "loves outdoors," "slob," etc.) or quite restrictive (as in *Prison Architect*, there is mostly one personality trait associated with each warden, such as "calm" or "lobbyist").

The ontological identity is close to what Kjevler calls the "playable character," which is the fictional character with her personality and background, to distinguish it from the avatar, understood here as the mechanism of fictional agency (Kjevler 2008).

Existential identity: The identity that the player builds for herself through decisions and actions in the game. It is partly derived from the ontological identity, as the latter designs different behavior patterns and possible actions in at least three different ways, that will be illustrated with different examples taken from *The Sims,* considered by many as the ultimate life simulation game, and *Prison Architect,* a goal-oriented simulation game.

The first way in which ontological identity influences existential identity in video games is simply by allowing a specific action. Typically "frugal" sims have the ability to clip coupons from the newspapers, while nonfrugal sims don't even have that possibility.

The second way in which ontological identity influences existential identity is by giving more incentives to perform a certain type of action. Still in the *Sims,* sims who love the outdoors will get "happy mood" points for spending time outdoors, which will make it easier and more rewarding to have them perform outdoorsy activities, such as gardening or fishing.

The third way in which ontological identity influences existential identity is by changing the horizon of possibilities, sometimes the entire course of the game. Embodying "the pacifier" in *Prison Architect,* which "reduces the overall temperature of your prison, making your inmates less likely to cause trouble," means that the player most likely will have to face extremely violent riot much less often, thus changing radically her game experience. If choosing a warden is an explicit decision to make that does not involve any rush, sometimes a trivial decision made in a hurry impacts massively the following of the game, such as in the action role-playing game *The Witcher.*

The Bogeyman identity: I argue that video games offer a fourth type of identity, the bogeyman identity, that is, embodying, in a constrained time-space, consequence-free framework, the identity of the person one does not want to be, but perhaps is attracted to. Rather than just fantasizing about being that care-free, or slob, or evil-spirited, or profit-driven person, the player can act like one in the separated time-space that video games offer.

In most games, the player then has to deal with the consequences of their actions and sometimes will have to start a new game altogether if the interactions with the nonvirtual others have become too aggressive or out-of-hand.

The one noteworthy exception is obviously *Grand Theft Auto* (GTA), in which there is no consequence for murder, rape, or any usually morally reprehensible action. GTA offers a space of total freedom, that is not so much immoral as amoral. The bogeyman identity is to video games such as GTA what *catharsis* is to classical theater.

> "Even in the case of those who treat a game purely as an ethical 'sandbox' to try out a variety of styles, the player still wants a sense of the consequences and the results because that's part of the pleasure and the appeals of these dilemmas" (Schreiber, Cash, and Link 2010, 76).

Not all games allow players to embrace all these types of identities. While existential identity is a common feature of most video games (since having to make decisions is at the core of video games), phenomenological identity is only relevant in video games that do display an avatar. As for ontological identity, I would like to suggest that the player can choose an ontological identity even in the absence of any avatar or explicit personality trait: somehow a hidden *intentional existential identity* that will influence her decisions in the game, but is not a design part of the game play and perhaps will not show on any indicator.

For instance, a player can start a game of *Civilization* having as a personal objective to be nonviolent, that is to never declare war, and in the case of other civilizations declaring war to her, try to cease all conflict as soon as possible. The player can even choose to embody a leader with advanced military traits and not use it as an advantage. This shows a possible dichotomy between the explicit, *game-designed existential identity* and the self-claimed, somehow secret *intentional existential identity*. Some scholars, adopting a psychoanalytic perspective, have argued that what happens in *Civilization* is an "intellectualisation" of brutal violence, "a transfiguration of primary sadism into strategic management" (Garandel 2012, 156). Beyond sublimation, what makes video games like *Civilization* a unique experience, though, is the "technical conversion of violence," in the sense that the epistemic value of the quest is subordinated to its pragmatic value: to win a game (Garandel 2012, 158).

This dichotomy between existential and intentional identity can only exist, however, if the player frees herself from the indicators' logics. Indeed, and as duly noted by French philosopher Mathieu Triclot, in video games such as *The Sims*, "the individual exists... under the pure form of a set of indicators. *The Sims* do in small what computing does at a larger scale: reduce a situation to its symbolic coordinates and handle it from a distance by acting on available information" (Triclot 2011, 216, *my translation*). If indeed the existential identity can mostly be reduced to a set of indicators, the intentional existential identity, by freeing oneself from the constraint of winning (and thus filling the indicators), can explore new combinations, combinations that perhaps will be valuable learning experiences. For instance, in *The Sims*, a player can choose to combine personality traits that will not help at all to achieve her avatar's lifetime wish, such as aiming to be a "heartbreaker" (have many romantic relationships), while having no sense of humor, being a workaholic and a snob (that trait usually keeps other sims away as the snob sim mostly talks about self). No doubt that this particular sim will be quite conflicted, and the player may have a hard time keeping it happy—that is, monitoring its indicators in a satisfactory way. Still, it may be an enjoyable game experience for the player, which reinforces her sense of self and her *existential identity* and *ontological identity* beyond built-in game options.

In any case, I would like to suggest that these multiple identities are, in themselves, a valuable ethical experience. The well-known economist and philosopher Amartya Sen has developed the idea that the self-awareness of belonging to multiple communities is a powerful step against violence (Sen 2007). Or said otherwise that the illusion of unique identity (being only or mostly an African American, a Hindu, a soldier, a gay person, a Palestinian, etc.) is one of the main causes of violence. To the heated debate about whether video games make players violent, I would like therefore to suggest that they can do the absolute opposite and foster a climate of nonviolence by making players aware of their multiple identities, actual and imaginary, as well as by allowing them to differentiate between their phenomenological, ontological, and existential identity.

Video Games as a Safe Learning Environment

In the second part of this chapter, I would like to argue that this opportunity for multiple identities is enhanced by the indefinitely iterative feature of video games that makes them a safe learning environment, desacralizing mistakes and providing a learning curve. The analysis of different backup systems will also provide a basis for discussion of Bernard William's concept of moral luck. Finally, the role of video games in controlling one's emotions will be considered.

An Iterative Learning Environment

The vast majority of video games offer the possibility to save a game through checkpoints, autosavings, or mindful backups (when the player chooses when to save). Autosavings are automatic and occur either on a time-scale (for instance, every fifteen minutes) or in case of turn-based video games, every n turns. Checkpoints on the other side are chosen by the game's developers as significant points to stop, typically after a specific challenge: if you have managed to pass it once, you do not want to try again. Checkpoints are therefore essential in a wide array of games (especially arcade games), and their relevance is paramount: "Bad checkpoints kill any and all interest I have in games, and replaying large sections of a level due to the developer's inability to use them well is infuriating" (Ben Kuchera, Polygon.com, February 25, 2014).

Finally, and usually compatible with either checkpoints or autosavings, the player has the possibility to perform mindful backups. Let alone specific technical issues (*I know the game usually crashes at this point so I always save my game right before*), mindful backups are a great way to enhance one's humility. Indeed, the possibility of mindful backup forces the player to anticipate *what could go wrong* at any time, and one's possible mistakes or shortcomings. Let's consider the following fictional monologue of a *Civilization* player:

> *I am considering declaring war on my expansionist neighbor. I have checked his military (he's weaker) and my finances (I'm wealthier), I have the diplomatic*

support of at least two other major nations, and I am technologically more advanced: what could go wrong? Oh well, let's save the game before I do declare war, you never know …

The possibility of autosaving, that is, the possibility to go back to this exact point in the game, before I declare war, in case my planning was wrong (*oops, he was researching Nuclear Power and I wasn't aware!*) invites the player to always consider the possibility of her failure. And if one's ego or hubris is too strong to practice mindful backups in this way, a hundred repeated failures in this safe learning environment usually do the trick. Regular gamers confess sometimes having the "mindful backup" automatism when interacting in real life: "*I was about to tell her my way of thinking; then I realized I couldn't do 'Ctrl + S' in case it would go wrong, and who am I to decide it* couldn't *go wrong?*" The infinite possibility of mindful backups in the universe of video games reinforces the perception that decisions and social interactions' consequences are irreversible in real life. This specific kind of humility learning is therefore transposable.

Consequently, video games in which mindful backups are possible allow for a type of mastery in reaching "the golden mean": if courage is somewhere halfway between cowardice and recklessness, as the Aristotelian view puts it, the iterative aspect and the backup system of video games allow gamers to actually gain courage by shying away from both extremes.

Video Games and Moral Luck

There is also another aspect made possible by the iterative aspect of video games: the great number of decisions and acts, condensed in few hours of gameplay, allow players to embrace the notion of moral luck as developed by Bernard Williams, without trivializing morality. Bernard Williams argues that although the very conceptual basis of our morality is meant to be immune to luck (according to Kant and followers, moral judgments ought not to depend on luck), there is in fact a significant part of luck in the way we perceive and assess the morality of actions (Williams 1981). A typical example is that of a person driving while being slightly drunk. If nothing special happens on the road, we may judge this person mildly irresponsible, but the condemnation will fade off pretty quickly. However, if she has an accident and kills someone (external circumstances), she becomes a reckless murderer. Another example used by Williams is that of an artist who isolates herself from other peoples' lives and needs in order to live her passion. If the artist becomes successful (external factor), she may become entitled in this radical choice. If she does not, she may be perceived as not only selfish and worthless but having taken a decision for which the scope of moral justifications (create something worth being prioritized) no longer exists. William's thesis has profound consequences on the very definition of morality, since it stops belonging to a transcendent category and instead becomes contingent.

Based on William's thesis on moral luck, Nagel's work later identifies four types of moral luck:

> **Resultant Luck:** The luck involved "in the way one's actions and projects turn out."
> **Circumstantial Luck:** The luck involved in "the kind of problems and situations one faces"
> **Causal Luck:** The luck involved "in how one is determined by antecedent circumstances."
> **Constitutive Luck:** The luck involved in one's having the "inclinations, capacities and temperament" that one does (Statman 1993, 60).

I argue that the infinitely iterative nature of video games may help players grasp the concept of moral luck in these various types, without narrowing morality to a trivial throw of dice. Indeed, luck is a characteristic paramount in the gameplay of many games. Sometimes, some level of information is available; sometimes, the player can directly impact luck by using some kinds of buffs (food, equipment, etc.). As an illustration, here are the messages a player can receive every morning while turning on the TV before starting her day in the simulation game *Stardew Valley* (a disciple, if not a spin-off, of *Harvest Moon*), with their luck coefficient.

The luck here is both circumstantial and resultant, if using Nagel's terminology: the luck affects both the type of situations the player faces, circumstantial luck (for instance chance of getting coal from rocks), and the luck involved in how to deal with this situation, resultant luck (for instance, the chance of discovering ladders while mining). So on an unlucky day, if a fellow villager asks for a mission in the mines ("Help wanted" missions), and if the player has chosen to embody a character with a helpful personality who always tries to complete this kind of quest, the player knows it is unlikely her character may fulfill the mission, independently of her previous experience in the mines (causal luck) or even the player's experience and will (constitutive luck).

Message	Luck coefficient
The spirits are very happy today! They will do their best to shower everyone with good fortune	+0.07 to +0.1
The spirits are in good humor today. I think you'll have a little extra luck.	+0.02 to +0.07
The spirits feel neutral today. The day is in your hands.	−0.02 to +0.02
This is rare. The spirits feel absolutely neutral today.	0
The spirits are somewhat annoyed/mildly perturbed today. Luck will not be on your side.	0.07 to −0.0
The spirits are very displeased today. They will do their best to make your life difficult.	−0.07 to −0.1

Through the backup system previously analyzed, video games offer this unique opportunity to live twice the same day, just with different luck coefficients, hence making empirically tangible the impact of various types of luck on the outcomes of a character's decisions. This diversity of experiences builds an understanding of ethics as a complex realm in which individual decisions are always to be measured against the messiness of life, hence allowing for a nonbinary understanding of ethical situations, in which moral luck is not an excuse for wrongdoing but rather a well-understood part to take into account in decision making.

Emotional Learning and Fostering a Climate of Benevolence

When a player experiences aggressive thoughts toward a virtual other, she knows that the artificial intelligence behind the game has not actually intended to hurt her or offend her. Video games can then be said to embody the popular wisdom known as "Toltec Lessons," one of the "agreements" reading: never forget that even when someone tries to kill you, it is never about you and always about him.

Therefore, video games offer the unique and valuable experience of having an actual emotion, with all its phenomenological characteristics (annoyance, anger, or frustration that can be translated into rising heartbeat, sweating, or extra focusing abilities) but without it being directed toward another living person. In other words: there is no one to blame, no one to direct one's anger to, no one from whom to assume evil intentions. Controlling one's emotions in this context is not an act of surrendering or a sign of cowardice, but of wisdom.

Finally, I want to suggest that single-player video games can foster both a demanding culture of self-control and a benevolent climate for other's ethical weaknesses.

Indeed, harshness in relation to others' weaknesses often arises from the strength of the conviction that "one would have done differently," that "one would have been capable of controlling oneself," and so on. By being confronted by many different contexts, by having to make a lot of decisions in unfamiliar environments, by being pushed to embody various identities, I believe single-player video games will develop virtues of not only leniency but benevolence toward others' weaknesses and ethical failures, because they will be more willing to imagine that this could have been them.

This is not to say that such a video games culture would be unable of having strong moral standards and of enforcing them, but rather that players experiencing Arendt's thesis of the banality of evil on a regular basis on their computer screens would be perhaps less prone to put away wrongdoers in the monsters box. Perhaps, having played in this ethical sandbox and recognizing a possible self in a virtual world, they would be more willing to engage in a highly necessary dialogue.

Transfer of ethical skills gained in video games into the "real world" is no easy claim. However, this is what many players' feedbacks suggest about some particularly engaging games. Among these, *This War of Mine* stands out. It is a war survival video game that was launched in 2014 and is very peculiar compared to most war-themed video games in that it focuses on the civilian experience of surviving through a siege, rather than frontline combat. The developers of 11-bit studios have shared that the game was inspired by the Bosnian War that they have experienced personally, and more specifically the Siege of Sarajevo. The gameplay looks like a tragic and stressful sims: the player controls characters who are most of the time tired, hungry, wounded, sick, and/or depressed, and who die regularly, without any hope of having them back. Indeed, *This War of Mine* (TWOM) does not allow for mindful backups: if your character is killed, there is no way to go back and try again—just like in real life. This is actually a feature quite common in survival games, and even to some extent in simulation games, that forces the player to experience a different kind of humility than the one described for games allowing for mindful backups. How is humility at play in a game like TWOM? At night one survivor scavenges in places that are more or less dangerous and tries to bring back food, meds, or construction materials. If an army guard sees him, or even another civilian, he might get killed. But if he doesn't bring back any food or meds, they might all die the next day anyway. There is no strategy here. One can never quite hoard nor plan. There are dilemmas, fueled by uncertainty. A second of recklessness, and the scavenger dies. But too much circumspection and it may be the entire shelter who vanishes, from lack of resources. Again, the golden mean is the way here, and the player learns it the brutal way.

But can these lessons really be passed on into real-life instances? Let us hear some players' comments on the topic (highlights by the author):

> Once again, this game shows you how war isn't anything like the first person/third person shooters we play. It's literally hell on Earth. **It really teaches you how to be humble** and appreciate what you have IRL. **And to think people like Syrians and the Rohingya are going through this in real life**. (Review posted on November 14th, 2017, on the Steam page "This War of Mine: Stories—Season Pass" by user KMZ 809—retrieved February 26, 2018)
>
> **I felt shame** when I went to the ruined villa and stole things that didn't belong to me, shame that deepened when I got back home to my own shelter and found that my mates had been wounded in another raid. The bandages and medicine I'd hoped to snatch from the other location? I never even found them… There was the time that the military came to my house looking for persons of interest in my neighborhood who'd taken food from an air drop. They said that the food was property of the state and they offered me cigarettes, canned food and fresh water in exchange for any info. Faced with three people who were either in "hungry" or "very hungry" states, I ratted out the neighbor. Despicable? Certainly, but war does that to a person… **I was aghast at how quickly my empathy eroded in a video game, which made me more cognizant of its fragility in real life. It's the kind of game that could potentially change**

the way you watch the news, treat others or cast a vote in an election. (Review posted on the Kotaku website, on November 18th, 2014, by Evan Narcisse: https://kotaku.com/this-war-of-mine-the-kotaku-review-1660267338—retrieved February 26, 2018)

Humility, shame, empathy—these reviews confirm, if needed, that dealing with nonhuman virtual others is not in any way preventing the players from developing feelings and emotions, as the case of Amazon's Alexa (Michelfelder 2020) also shows. Even more, the wide range of human emotions at stake seems not only to be unavoidable within the video game experience but to transfer to real-life thinking quite easily. If video games offer a safe learning space in which emotions can thrive and virtues develop, these are by no means confined behind a screen.

References

Aarseth, E.J. 1997. *Cybertext: Perspectives on Ergodic Literature*. JHU Press.
Cogburn, J. and M. Silcox. 2008. *Philosophy through Video Games*. New York, Routledge: Taylor & Francis.
Garandel, P. 2012. "Malaise dans Civilization." In *Les jeux vidéo comme objet de recherche*, edited by Samuel Rufat (Sous la direction de), Hovig Ter Minassian (Sous la direction de), 149–175. Paris: L>P Questions Théoriques.
Geary, W.T. and R.R. Sims. 1994. "Can Ethics Be Learned?" *Accounting Education* 3 (1): 3–18. https://doi.org/10.1080/09639289400000002.
Gibert, M. 2015. *L'imagination en morale*. Paris: L'avocat du diable.
Johnson, S. 2006. *Everything Bad Is Good for You: How Popular Culture Is Making Us Smarter*. Penguin UK.
Kjevler, R. 2008. *What Is the Avatar* (Unpublished thesis). University of Bergen, Bergen.
Lafrance, J.-P. 2006. Les jeux vidéo: à la recherche d'un monde meilleur. Paris: Lavoisier.
Langlois, L. and C. Lapointe. 2010. "Can Ethics Be Learned?: Results from a Three Year Action Research Project." *Journal of Educational Administration* 48 (2): 147–163. https://doi.org/10.1108/09578231011027824.
Michelfelder, D. 2020. "Relating to Alexa Relating to Us." In *Relating to Things Relating to Us*. This volume.
Ogien, R. 2007. L'éthique aujourd'hui: maximalistes et minimalistes. Gallimard.
Pohl, K. 2008. "Ethical Reflection and Emotional Involvement in Computer Games." In *Conference Proceedings of the Philosophy of Computer Games*, edited by S. Günzel, M. Liebe, and D. Mersch, 92–107. Postdam: University Press.
Ritter, B. A. 2006. "Can Business Ethics Be Trained? A Study of the Ethical Decision-Making Process in Business Students." *Journal of Business Ethics* 68 (2): 153–164. https://doi.org/10.1007/s10551-006-9062-0.
Roeser, S. 2010. *Moral Emotions and Intuitions*. Springer.
Schreiber, I., B. Cash, and H. Link. 2010. "Ethical Dilemmas in Gameplay: Choosing Between Right and Right." In *Designing Games for Ethics: Models, Techniques and Frameworks*, edited by K. Schrier, 72–82. Hershey, PA: Information Science Reference.
Sen, A. 2007. *Identity and Violence: The Illusion of Destiny*. Penguin Books India.
Sicart, M. 2005. "Game, Player, Ethics: A Virtue Ethics Approach to Computer Games." *International Review of Information Ethics* 4 (12): 13–18.
Sicart, M. 2011. *The Ethics of Computer Games*. Cambridge, MA: MIT Press.

Statman, D. 1993. *Moral Luck*. Albany, NY: SUNY Press.
Steinbock, A. J. 2014. *Moral Emotions: Reclaiming the Evidence of the Heart*. Evanston, IL: Northwestern University Press.
Švelch, J. 2010. "The Good, the Bad, and the Player: The Challenges to Moral Engagement in Single-Player Avatar-Based Video Games." In *Ethics and Game Design: Teaching Values Through Play*, edited by K. Schrier, 52–68. Hershey, PA: Information Science Reference.
Triclot, M. 2011. *Philosophie des jeux vidéo*. Paris: Editions la Découverte.
Williams, B. 1981. *Moral Luck: Philosophical Papers 1973–1980*. Cambridge: Cambridge University Press.

6 CASTING THINGS AS PARTNERS IN DESIGN: TOWARD A MORE-THAN-HUMAN DESIGN PRACTICE

Elisa Giaccardi

Introduction

Understanding how we live with things, and in turn, how things come to live with us calls for methodologies that go beyond a focus on humans. As scholars, from material culture studies to object-oriented philosophies, we have come to appreciate the agency and social life of the things we make. Yet as designers, we fail to move past the blind spots of our intentions and give things a voice in doing design work. We still believe that the relationship between humans and things is unidirectional: only humans make things.[1]

In the face of such negligence, design is rapidly being widened and disrupted by the flood of data technologies under the name of Internet of Things, machine learning, and artificial intelligence. These technologies have a profound effect on the nature of products and services, enabling things to "make" things too through the exchange and processing of data (e.g., generating playlists, delegating assistance requests, arranging smart contracts). This raises urgent and fundamental questions about the way designers will participate in this expanded world of design next to the things enabled and made autonomous by data technology.

The work presented in this chapter suggests that things—as they begin to be artificially enabled to sense and perform autonomously by means of software, sensors, and actuators—may have access to perspectives and fields that we as humans do not. My argument is that casting such things as partners in design may in turn enable designers to explore new spaces and objects of design. More precisely, bringing a thing perspective to design provides a different point of view that can help us see what is not immediately apparent to human observation (because on a different perceptual scale) but also what may fall outside of our sense of relevance (because not yet accounted for). This can help problematize what we take for granted and offer different ways of understanding what we know and what we do, humans and nonhumans alike.

So the question addressed in this chapter is not so much whether things have their own intentionality, ontologically speaking, and whether this is manifest or opaque. The question is methodological. It concerns the co-ability of things to make things next to us in ways that are uniquely artificial, and the role they can play in the work of doing design.

But designers need to creatively and extensively exercise and practice the principles of a new approach, and to take the underpinning technology seriously, before they can actually design with it (Giaccardi, Speed, and Netten 2016). And so this chapter unpacks, by means of case studies and concrete examples extending over an arch of three years, how designers and things worked together as partners and the trade-offs of more-than-human design practices.

Data Technologies and the Agency of Things

Before moving to the case studies, though, it is important to clarify what is meant by "thing" in this chapter, and what is unique about data technologies from a thing perspective.

As argued in design research after Heidegger[2] (Tonkinwise 2005) and Latour[3] (Ehn 2008; De Michelis et al. 2011), designers make things. A "thing" is not the artifact in its straight materiality but a nexus of relations that has the ability to shape ways of doing and open up new futures. In simple terms, we could think of a "thing" as the design artifact(s) plus the people (or other artifacts) that relate to it plus how they relate to it. In design, we often think of this relation as one of use, though of course "use" is a simplification of the more entangled relation we have with things, and things with us (or with other things) (Redström and Wiltse 2018).

Today autonomous vehicles; assistants such as Alexa, Google Home, and Cortana; drones that deliver purchases within minutes of placing an order; Ethereum tokens; and smart contracts are new kind of things that increasingly do business with humans and with each other (Iqbal 2018). As things become enabled through the exchange of data to see, make judgments, and perform actions that create new connections to and shape new relations with both humans and other things, we must acknowledge that increasingly also things make things.[4]

This problematizes how things take part in design next to us, professional designers and everyday designers alike, and how their uniquely artificial competence and skills, their point of view, can be brought to bear on design work in ways that broaden and balance both human and nonhuman perspectives. In a world where the complexity and scale of design problems have grown, and where distinctions between design and use, subject and object, producer and produced have blurred, the challenge of design is not a matter of getting rid of the emergent and placing the human more firmly at the center. It is rather a matter of how to partner with things in doing design and make it an opportunity for more creative and hopefully more appropriate solutions.

Attributing agency to things is not a new concept (Brown 2001). Actor network theorists discuss the ontological symmetry of humans and nonhumans, in which material forms take on the characteristics of humans: they judge, form networks, speak, and work performatively (Engeström and Blackler 2005). Similarly, anthropologists concerned with materiality have suggested that objects are dynamic and emergent entities that contain their own life forces, energies, and histories (Appadurai 1986; Miller 2009; Hodder 2012; Gatt and Ingold 2013). More recently, object-oriented philosophy posits that things do not exist just for us (Bogost 2012); they can be many and various (Bryant 2011), but no matter their size, scale, or order, they enjoy equal being (Harman 2009). Though criticized by object-oriented philosophy for disavowing any reality external to human experience, postphenomenology too considers things' agency (or more precisely mediation) as potentially withdrawing from human understanding and perception, hiding, receding into the background of human awareness even when in use (Ihde 1993; Verbeek 2005).[5] In design, the gap between things and us is often addressed through the speculative exploration of the new forms of attachments people may develop toward things despite the gap (Di Salvo and Lukens 2011; Wakkary, Houser, and Oogjes, this volume). As shown in development psychology (Piaget 1959) and the psychodynamic tradition in psychoanalysis (Turkle 2007), this often leads to people ascribing intentionality and consciousness to inanimate objects (McVeigh-Schultz, Stein, Watson, and Fisher 2012; Marenko 2014; Rozendaal 2016), seeking similarities between the animate and the inanimate (Giaccardi, Speed, Grossen, and van Allen 2014).

This ontological gap and the design implications of the perceived intentionality of things for user experience are not the concern of this chapter. As argued in a previous publication (Giaccardi 2019), data technologies challenge design practice to respond to three emerging shifts: the *agential shift* toward the inclusion of things in design as partners, the *temporal shift* toward always-available opportunities for co-creation, and the *material shift* toward more infrastructural and fluid forms of generating and sustaining value. According to Tonkinwise (2015), this move in technology development toward platforms that bring people closer to the production and distribution of products and services had already been anticipated by metadesign (Giaccardi 2005; Busbea 2009) and postindustrial design (Cross 1981; Hunt 2005).

In the attempt to further unpack the agential shift brought about by data technologies and its implications for design practice, this chapter suggests that a useful perspective to position things in design is to consider agency in terms of their capability to co-perform, to carry out artificial performances next to people (Kuijer and Giaccardi 2018). Over the course of repeated performances, and alongside newly developed artifacts with unprecedented capabilities such as machine learning algorithms, human and artificial minds and bodies change and learn to take on different roles in co-performance. The idea that data technologies enable things to perform according to skills different than humans had already been suggested in an earlier work (Giaccardi, Speed, Cila, and Caldwell 2016). The elaboration of this idea as a modification of theories of practice, in terms of co-performance, resonates with feminist reconceptualization of performativity, according to which agency is not something that people or artifacts have; it is the emergent result of how the world actively and continuously configures and reconfigures itself (Barad 2003; Bennett and Joyce 2013).

Considering data-enabled, algorithmic things as capable of performing practices next to people challenges and offsets the idea of humans and nonhumans as independent from each other and autonomous. Even further, considering things as capable—as I argue in this chapter—of actually "making" things next to people (because of their ability to co-perform) shifts the locus of doing design toward a fundamentally participative relation, one that is informed by capabilities and doings uniquely human and uniquely artificial.

Things as Design Partners

In unfolding a future in which algorithms and autonomous devices increasingly make things together with humans, it is imperative to move past the blind spots and unilateral arrangements of human-centered design. The idea of things as partners in design, as presented in this chapter and further elaborated through the case studies, builds on the conceptualization of the co-ability of things to make things next to us in ways that are uniquely artificial. As argued before, things make things too.

Bringing a thing perspective to design offers an alternative that harbors different ideas about human and artificial expertise and skills, and their relation. The patterns revealed through a thing perspective emerge at the intersection of the data and trajectories that things give access to and the inquiry that humans bring to it. This is not done simply to provide different and unique information about people or to offer out-of-the-box inspiration for original solutions. The aim of a thing perspective is fundamentally to problematize and enable the exploration of spaces and objects of design that are not constituted yet but emerge in response to nonhuman perspectives. As beautifully captured by Tim Ingold: "It is not, then, that things have agency; rather they are actively present in their doing... And as things carry on together, and answer to one another, they do not so much interact as correspond. Interaction is the dynamic of the assemblage, where things are joined up. But correspondence is a joining with; it is not additive but contrapuntal, not 'and ... and ... and' but 'with ... with ... with'" (2017, 13).

Considering this co-performative relation as design partnership—as the co-dependent ability to make things—moves us past the limitations of using the notion of co-performance to examine and predict analytically the practices performed by autonomous devices after design (Kuijer 2018)[6]. And it allows us to conceive of the artificial performances of things as taking part in a fluid and unstable more-than-human design practice, which is not separate from professional or everyday design practices but entangled with them in the looping and blurring of design time and use time.

But what are these more-than-human design partnerships for? By exerting the ability to access trajectories unattainable to human observation and make design proposals, potentially contesting our worldview, things contribute a different perspective and unique insights that enhance, complicate, and even challenge those of humans (Giaccardi, Speed, Cila, and Caldwell 2016). Considering things as design partners is different from looking at them as collaborators in achieving human originating purposes (Kaptelinin and Nardi 2006; Rozendaal 2016; Grudin 2017). But a partnership with things requires engagement, and practice. It assumes to "spend time" with things and "work together with" them. It requires sustaining collaborative processes with things and among things that offer different ways of understanding what we know and what we do (Gunn and Donovan 2012).

Different ways of understanding allow for reframing and reconfiguring social and material relations, and are inherently performative and transformative. Engaging with things for an extended period of time and reflecting on what we usually take for granted open up and articulate spaces and objects of design that were previously unattainable. It troubles distinctions between subjects and objects, producers and produced, and in so doing supports ways of understanding and designing that take place *after* design (at use time), but also *with* and *beyond* the design work at project time.

Three Cases of More-than-Human Design Partnership

In this section, I describe three cases in which designers and things have worked together as partners and examine the emergent character of their partnership. A final reflection on what such partnerships may add to human-centered design practices will be offered at the end. The cases are based on projects conducted at the Connected Everyday Lab, Delft University of Technology between 2015 and 2018. The projects employed different techniques to include a thing perspective in the process of doing design, from life logging to bespoke sensors and open libraries for data visualization to expert machine learning.

Each case is unpacked along two axes. First, I examine what things "have seen": what trajectories and data worlds the things enlisted as partners in the project enabled the designer(s) to access through artificial sensing and data analytics. As introduced in earlier work (Giaccardi, Cila, Speed, and Caldwell 2016), the expression "data worlds" is used here instead of "data" to consider the arrangements among people and things, and things and things, and thus the ecosystems in which these are imbricated. It is access to these horizontal relations and arrangements, and the unique insights generated about such ecosystems, which things bring to bear on the design partnership through what we have referred to in other work as "thing ethnography" (Giaccardi, Cila, Speed, and Caldwell 2016). Then, I examine how a thing perspective may have "problematized" the design space: whether things unsettled the designer's assumptions; demonstrated the problem to be more uncertain, more nuanced, or more complex than originally assumed or regarded; and how a more-than-human partnership configured within the process of sensemaking and framing. "Framing" is used here to refer in design terms to the result of "sensemaking," that is, to the outcome of the "constant process of acquisition, reflection, and action," which is fundamentally based on the perspective or point of view of those participating in doing design (Kolko 2010). In design research, framing is conceptualized as the hypothetical way of looking at the problem (Dorst 2015a, 25), the talking into existence of "assumed-to-be-real facts" or "facets of things" that do not exist yet (Kolko 2010). As such, framing configures the scope of design work (Kolko 2010) and can mutate what was initially envisioned as a desired outcome (Dorst 2015a), helping to think of problems (and thus solutions) in always new ways. Ideally, in a world of increasing complexity and blurring of the design disciplines, the creation of frames should painstakingly embrace the "unknown nature of the outcome" (Dorst 2015a). Attributed conventionally to expert design practice, sensemaking and framing are opened up in this chapter to the perspective of things and their performances. It is through more-than-human sensemaking and framing that design achieves its intention.

Lastly, I will briefly mention the objects of design that emerged out of the partnership with things within each project, ranging from speculative demonstrators to product concepts.

Envisioning Culturally Sensitive Innovation for Taiwanese Smart Mobility

Taiwanese use scooters to carry out many of their daily activities, especially in highly populated urban environments. It is a complex relationship that Taiwanese have with their scooter in everyday life (Lin 1998; Lai 2010). Scooters are not just a means of transport but intimate companions of daily practices, from shopping at the street market to transporting goods and picking up children. Scooters impact how Taiwanese perceive and imagine the world.[7] Blind to this reality, smart mobility has so far mostly focused on energy consumption and efficiency optimization.[8] In the thing ethnography that we conducted on the use of scooters in Taipei as part of a project in collaboration with the National Taiwan University of Science and Technology,[9] we focused instead on revealing the imbricated web of relations that develop around scooters in Taiwanese everyday life. Our goal was to envision culturally sensitive forms of smart mobility for the Taiwanese context.

Trajectories and Data Worlds Accessed through Artificial Sensing and Data Analytics

Six scooters were equipped with intelligent cameras and sensors and enlisted as partners to collect data in the field and help generate insights from a thing perspective. By enabling scooters to artificially "sense" and "see," we hoped to gain insight into their social life, and specifically into the cultural idiosyncrasies of the relationship that develops between scooters and Taiwanese in everyday life. In order to access this data world, we instrumented scooters for artificial sensing with a time-lapse camera and a repurposed smartphone for recording GPS tracks, for a data collection period of three days. The time-lapse camera was attached to the scooter's handle, facing the scooterists, and was set to take a photo every ten seconds. Without the shutter button being controlled by a human, as shown in early studies (Giaccardi, Cila, Speed, and Caldwell 2016), the time-lapse camera would have captured both recognizable and hidden social practices around the scooter and the broader systems of relations of the scooter with other things. The smartphone with the app for recording GPS tracks was used instead to collect data on the scooter's daily trajectories and other geographical data such as location and acceleration. As opposed to the domestic objects instrumented in earlier work (Giaccardi, Cila, Speed, and Caldwell 2016), the scooter is highly mobile and may reach a speed that hinders the efficacy of lifelogging techniques. Finding the right placement for the camera was tricky on a scooter. But by complementing lifelogging with geographical data, we enabled the scooter to "see" and "sense" also the more dynamic and eluding elements of its relationship with Taiwanese everyday life.

Knowing that scooters in Taiwan are used very differently depending on the lifestyle of their owners (Lin, 1998; Lai 2010), we then invited six people with different jobs and different lifestyles to take part in the thing ethnography with their own scooter (Figure 6.1). Human participants included: a student, an office worker, a motorcycle enthusiast, a housewife, an insurance agent, and a plumber. Nonhuman participants included: Pudding, Jog100, Moon, Fighter125, Breeze125, Vino50.

Once data were collected, they were organized and presented in a format conducive to role-playing. The goal was to limit the efforts needed to approach data as an analyst and facilitate instead immersion into the social life of the scooter. Six well-trained professional actors were invited to engage with the data, role-play one of the six scooters, and "speak" on its behalf when interviewed by the designer. The rationale for this choice was to develop a technique that would help the human designer empathize with the always-withdrawing inner life of things. The assumption was that professional actors are well positioned to speak for the nonhuman, as they are trained to bring people and things to "life" in a highly relatable way. The technique we invented is called Interview with Things (Chang, Giaccardi, Chen, and Liang 2016).

Pudding and Mrs. Cheng.

Jog100, Mr. Lin and Mrs. Wu.

Moon and Mr. Chen.

Fighter125 and Mr. Liu.

FIGURE 6.1 Portrait photos of participating scooters and scooterists. Photos by Wen-Wei Chang.

Videos edited from the same perspective as the time-lapse photos were one of the formats in which data were presented to professional actors. These videos, recorded from the point of view of the scooter, were particularly useful to help actors immerse themselves in the dynamic experience of being a scooter. Actors showed great potential to decenter human perspective. During the interviews, the actors not only "felt what the scooter felt" but "thought and reflected in a scooter's way." For example, one actor implied her difficulty understanding some "too human" words. Another actor also mentioned that, for her as a scooter, all things can be categorized into things that don't move (e.g., buildings), things that move by themselves (e.g., humans and street dogs), and things moved by other things (e.g., scooters and cars). In all these cases, the actors skillfully immersed themselves into being a thing, bracketing their human-oriented way of thinking. By thinking and reacting as scooters, actors helped broaden our understanding and imagination about the scooter as a thing.

One of the insights generated from a thing perspective, for example, was that the function of the scooter is dependent on the speed of the scooter. Depending on their speed, scooters in Taipei become carts for grocery shopping at the street market, temporary addresses to which to deliver flowers, or benches for chatting with your friends in the parking lot. In the ideation phase this contributed to the speculative concept where the scooter's tailpipe, when used at full speed, is used as a heater for a variety of improvised uses on the road, from ironing to warming up foods and drinks, which work and make sense within the social and cultural norms of Taiwanese everyday life.

More-than-Human Sensemaking and Framing

Scooters are notoriously low cost and easy to modify. As described in studies of material culture (Lin 1998; Hebdige 2001; Lai 2010), scooters are often subject to creative appropriation along their material and functional dimension. The handles can be arranged as a rack to hold drinks, and the backseat can be used as storage to contain goods. Designers have learned from this and incorporated some of these elements in the scooter as product features. However, the thing perspective brought to bear on the exploration of a possible design space for smart mobility has revealed that there is more to the arrangements between Taiwanese, scooters, and environments than material and functional factors. The unique social relations and meanings that develop between scooter and scooterist are an equally important element in Taiwanese everyday life. In other words, Taiwanese value scooters not only because of their usefulness for commuting and delivering goods but also because of the diverse and dynamic meanings the scooter acquires in the way it helps build and maintain social connections. The tension between the intent and expectations of the designer (focused on material and functional features) and those of the user (concerned instead with the social quality of the scooter) was expressed several times by scooters during the interviews. For example, the small scooter revealed that while the "double (sometimes triple) carries" is not considered proper use by its designer, the social quality of physical proximity expressed by this misuse is instead highly valued by lovers and family members.

As social qualities and material affordances of a product (including its functionalities) go hand-in-hand, making sense of the collected data together with things in the interviews helped understand that the dynamic and unique relationship that develops between scooter and scooterist according to different usages and in different situations should be considered an important element of the socio-material arrangements that constitute Taiwanese's everyday life.[10] The importance of a scooter's social qualities and how these translate into usage, how social qualities, material affordances, and creative misuse are imbricated in the Taiwanese context through everyday practice, is ignored instead by mainstream smart mobility as currently framed.

Interviewing scooters was not only valuable to make sense of the collected data. It was also an inspiring intervention for the designer to speculate what is like to be a thing within specific socio-material arrangements. By encountering and empathizing with a convincing nonhuman actor, the designer gained rich and novel inspiration. The interview was not just a solo performance by the actor but a cooperative speculation by the designer and the "thing" (as enacted by the actor's performances). To help the actor understand the thing and decenter momentarily from a human perspective, the interviewer also needed to decenter his human-centered logic. For example, instead of using terms such as "personal relationship" in the interviews, he used the term "scooteral relationship" to help consider the scooter as a thing and not just a product. Through the interviews, the designer was able to defamiliarize and engage in an imaginative design partnership with the always-withdrawing nonhuman. By making sense and speculating through role-playing, the design technique used in this project was a sincere invitation for both humans and nonhumans to engage and understand each other and explore together culturally sensitive forms of smart mobility for the Taiwanese context.

Emergent Objects of Design

The outcomes of this project included a series of speculative scooter portraits and a speculative set of accessories for smart scooters. The scooter portraits were commissioned to an illustrator and directly based on the transcripts of the interviews with things conducted with the actors (Figure 6.2). These portraits served the designer as an intermediate object to organize insights and move toward his final concepts.

The final concepts were prototyped and exhibited in Taipei in 2017 as a speculative set of three accessories for smart scooters aimed to foreground scooter's social qualities in a playful manner: pipe heater, sound generator, and red light pointer & atmosphere meter.

FIGURE 6.2 Scooter portraits based on the transcripts of the "interview with things" conducted with professional actors. Image courtesy of Wen-Wei Chang.

Pipe heater (Figure 6.3) is an open-ended device for scooterists to reuse the heat produced during a ride, for example, preparing a warm lunch during a working commute. As a concept, the pipe heater broadens the margins of resourcefulness of the Taiwanese scooter in everyday life by inviting creative appropriations around one's own lifestyle and daily social interactions.

FIGURE 6.3 Pipe heater is the speculative concept for an open-ended device that reuses the heat produced during a ride for personally and socially meaningful activities (e.g., warming up food, sharing a hot drink). Photos by Wen-Wei Chang.

Sound generator (Figure 6.4) is a smart audio component producing a sound out of a scooter's engine that is personalized according to a scooterist's riding patterns, for example, an aggressive and high-pitched sound for a racer and a prim and proper sound for a gentle scooter rider. This concept enriches people's ability to express themselves through the scooter. In the final exhibition, five sound artists were invited to create unique sounds for the six scooters in the scooter interviews, to help the audience imagine how the generated sounds might sound like.

Red light pointer and *atmosphere meter* (Figure 6.5) are smart dashboard components designed to bring people physically closer. The red-light pointer is attached below the original speed meter pointer. Rather than telling the current speed, the red-light pointer indicates the "recommended" speed to encounter more traffic lights, and more opportunities of hard braking and physical contact. The atmosphere meter is a pointer attached below the original fuel meter pointer.

FIGURE 6.4 Sound generator is the speculative concept of a smart audio component for a scooter's engine that is personalized according to one's riding patterns and needs for social expression. Photos by Wen-Wei Chang.

FIGURE 6.5 Red light pointer and atmosphere meter are speculative concepts of smart dashboard components designed to bring people physically closer and create intimacy. Photos by Wen-Wei Chang.

This pointer does not indicate the amount of fuel in the tank but visualizes the current social atmosphere on the scooter. As a concept, these smart dashboard pointers aim to encourage and facilitate social and intimate interactions on the scooter.

Stimulating Creative Dialogues in Democratized Manufacturing

Over the past decade, technologies for computer-aided technical drawing (CAD) and rapid prototyping and manufacturing (CAM), in combination with online platforms for the distribution of creative projects such as Etsy and Instructables, have revived interest in self-made, do-it-yourself (DIY) products. The phenomenon referred to as the Maker Movement has promoted a further step towards a democratization of design, as technology effectively puts control over geometry, materiality, and assembly into the hands of a new pool of makers. However, design encompasses more than control over the object itself; it also involves an understanding of the object in its context. This type of appropriateness between an object and its context cannot always be engineered, because it varies in different situations of use and under different circumstances. Production tools alone are not sufficient for the democratization of design. The potential of modern-day DIY lies far beyond hundreds of differently styled iPhone cases to choose from. In this project,[11] we focused on how rapidly spreading Internet of Things systems for the home might help makers discover new applications of their crafting and making skills. Our goal was to understand how to introduce a thing perspective in the creative process of the DIY practitioner and help them open up their design space.

Trajectories and Data Worlds Accessed through Artificial Sensing and Data Analytics

The project enabled seven makers to deploy Wi-Fi-enabled sensor modules and conduct thing ethnographies of their homes, with the intent to learn more about their context and open up the design space for home improvement. Compared to the previous case, the challenge of this project was to work with nondesigners, unfamiliar with the general principles of contextual inquiry and primarily driven by the need to express themselves and learn new skills (Atkinson 2006; Kuznetsov and Paulos 2010). A total of seven makers took part in the project. In this chapter, I will discuss the second study of this project, which used bespoke sensors and open data visualization libraries.[12] In this study, we asked four makers to think of what they wanted to make, and in what ecology of other artifacts and practices their object would have ended up. Then, in discussion with participants, we hacked some of the artifacts in this ecology by enabling them to collect sensor data from a thing perspective and access their domestic data world. The assumption was that this would have revealed additional design opportunities. The study concluded with a creative session, in which the makers first discussed their process and findings and then were asked to swap their data and design for each other.

In a couple of cases, makers were able to come up with a good sensor placement, discover something they did not know, and generate solutions for preventing cooking smells or reminding partners to take their shoes off when working late at home. However, in trying to address nondesigners unfamiliar with sensors, the project failed to help makers partner with things to generate unexpected insights that could open up novel design spaces for home improvement. Relating graphed sensor data to real-life phenomena was problematic for our participants. Anticipating what kind of data they would receive from the sensors, and owning the process of which sensors to use and where to place them, was even more problematic. Equally complicated for the makers was to move past (or through) the sensitivity and sampling rate of the sensor. It was notable that most participants came up with design ideas that revolved around automation. Examples are curtains that open automatically when the sun rises, based on the light sensor, or an alarm that would sound when housemates would start cooking without opening the window. Instead of using things to access previously unattainable trajectories, the makers seemed eager to incorporate the sensors in their solutions, expressing a tendency to what Amram (2016) describes as "automation fixation." Makers experienced fixation also in relation to what to expect from the sensor data, which Amram (2016) refers to as "phenomenon fixation." For example, once the constant relation between room temperature and one of the maker's shoe cabinet was interpreted as signifying the presence of people in the room, the maker became blinded to every other possible meaning of the data.

However, in the concluding session in which makers were asked to swap data and design for each other, a different kind of sensemaking emerged. Whereas makers had a difficult time making sense of their own data and accessing the broader data world in which the domestic object under examination was (or could be) imbricated, swapping data and designing for each other took away these barriers by removing expectations and fixations. For example, in the case of the shoe cabinet, the constant temperature was not considered a failed measurement by the others. Instead, it gave way to the out-of-the-box idea of a terrarium where to farm reptiles for leather.

More-than-Human Sensemaking and Framing

While there is something to be said about the importance of technological literacy for being able to capitalize on artificial sensing and data analytics in a creative process, this project suggests that even in the case of nondesigners, it may be possible to find ways to partner with things. The project involved participants who varied in terms of their involvement and interest with the sensors, and our findings made clear that the more engaged and knowledgeable the maker was, the more he could engage a "creative dialogue" with things. However, it is in the concluding session during which makers collaborated to make sense of the data that they were able to use the results of their thing ethnographies to enhance communication and creativity within the convened DIY community and problematize the original scope of their projects. Questions like "Why did you put that sensor there?" proved to be excellent sensemaking starters.

In speaking of their placement, and making otherwise tacit design considerations explicit, things counteracted makers' fixations and frames of reference and revealed new design opportunities.

Emergent Objects of Design

MakeDo (Figure 6.6) is a speculative design concept for fostering creative dialogues between makers and things in democratized manufacturing created by Amram (2016) that embodies the findings of this research and promotes a distributed type of the design partnership with things. Elements of the design address recurring issues observed in makers' thing ethnographies. The main difficulty in casting things as partners in makers' workflow was that ethnography (and contextual inquiry more in general) is an unfamiliar component of the DIY practice. The observable consequences of this unfamiliarity are a fixation on automation projects and phenomenon fixation.

MakeDo can be shortly described as a platform for DIY recipes where data collected from things are an integral part of the making process. On the *MakeDo*

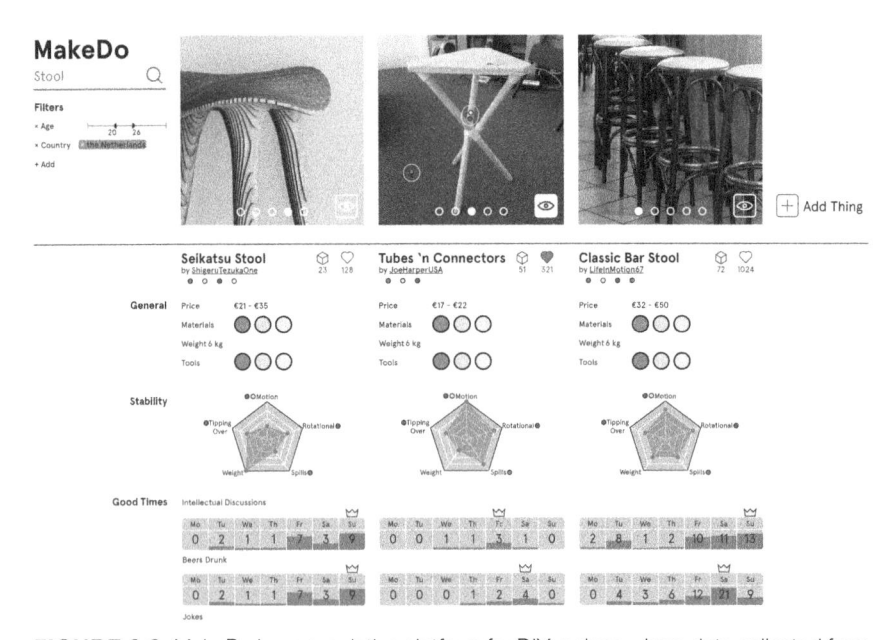

FIGURE 6.6 MakeDo is a speculative platform for DIY recipes where data collected from things are an integral part of the making process with the goal to foster creative dialogues between makers and things in democratized manufacturing. Image by Tal Amram.

platform, community members do not only share the "making" of an object (its DIY recipe) but also its "doing" (its data). Instead of a linear process from author to platform via the DIY recipe, recipes that are shared on *MakeDo* also include sensor data about the use of the thing generated and collected by its multiple physical instances. Conventionally, DIY recipes are created, published online, downloaded, made into a physical artifact, and eventually used. *MakeDo* closes this loop by feeding use data back into the recipe.

For data sharing to be so intimately interwoven into the making process, the design concept envisions radically compact and simple sensors called "knots" (Figure 6.7), which could be bought in the hardware store or ordered online to the exact specifications and quantities of the project the maker may be

FIGURE 6.7 Example of the executed DIY recipe of a stool with sensing knots from the MakeDo community. Photos by Tal Amram.

undertaking. As single knots are inserted into a physical artifact, they begin to form a small local network of cooperative sensors and exchange their data with the *MakeDo* platform.

Community members can then publish and share not only the blueprint of the DIY recipe, including placement of the knots, but also a plugin for data aggregation and visualization. Using different plugins, a maker could then compare several DIY recipes of stools based on requirements such as the measured stability or the amount of jokes inferred through data.

The feedback that the thing can now send to inform its own blueprint blurs the traditional dichotomy between design time and use time. This poses interesting opportunities for parametrically designed objects or procedurally generated designs. Collaborating algorithms from several plugins could map what design decision has what effect on the use of an artifact, and the resulting information will give makers new source of inspiration. This is the very essence of *MakeDo*: combining design time and use time into a cyclic process. There is no fixed optimum to strive toward (at least not enforced by the platform), but an endless string of discoveries to be made.

Empowering Older People to Age Resourcefully in the Connected Home

In products designed especially for older people, the inventiveness and resourcefulness of elderly are often underestimated in favor of designs mistakenly assuming older people to be helpless and frail (Giaccardi, Kuijer, and Neven 2016). In the Resourceful Ageing[13] project, we focused instead on what older people can still do and the strategies they put in place to creatively cope with their ageing skills. Our goal was to find out how to design connected products for elderly people that can improvise in use and thus remain appropriate to a large variety of situations.

A collaboration between industrial design, computer science, social sciences, and industry partners,[14] this project used a combination of machine learning and ethnographic fieldwork to research and prototype designs that can support the everyday practices of resourcefulness of elderly people.

Trajectories and Data Worlds Accessed through Artificial Sensing and Data Analytics

Because resourcefulness is a dispersed practice that is difficult for the human eye to observe and capture (Kuijer, Nicenboim, and Giaccardi 2017), we invited five households of people in between sixty-five and seventy-eight years of age as well as their domestic objects to take part in the thing ethnography. Human participants included four females and one male living independently at home, two of which with their spouses. Nonhuman participants included doors, fridges, chairs, and remote controls as well as unique "things," such as spider stick and rope on stairs.[15]

These were selected together with human participants through a combination of sensitization techniques and ethnographic fieldwork[16] (Figure 6.8).

We deployed a bespoke wireless sensor network infrastructure and instrumented with artificial sensing capabilities eight objects per household, for a total of thirty-two domestic objects. Over a period of two months, we collected 133 MB of sensor data from three of the participating households. Sensors sampled when objects moved in space as well as environmental data.[17] We then used unsupervised machine learning techniques to discover structure from data and assign meaning to it. The intention was to ask our nonhuman partners about routines developed within temporal patterns of day and night, weekday and weekend, which might suggest practices of resourcefulness too dispersed for a human observer (including our human participants) to discern.

The resourcefulness witnessed by objects in elderly homes and captured by human observation looked like the magnet in Figure 6.9: a thing, a very mundane entanglement, central to the resourcefulness of one of our elderly—who uses the magnet to keep together small objects she would not be able to grab when flat on the table.[18] But what the algorithms that we developed for this project were able to see is the probability of a thing being handled at a particular time of day and the clusters of things being handled at the same time: a hint to the possibility that these things may be used together often as part of a dispersed yet established practice of resourcefulness, which escapes human observation or normative sense of relevance. By moving from the analysis of raw temporal events (Figure 6.10) to the interpretation of their clustering at an abstract level (Figure 6.11), we did

FIGURE 6.8 Resourceful Ageing: Selecting nonhuman participants via a combination of sensitization techniques and ethnographic fieldwork. Photo by Iohanna Nicenboim.

FIGURE 6.9 Resourceful Ageing: Participating magnet, central to the resourcefulness of one of the human participants. Photo by Iohanna Nicenboim.

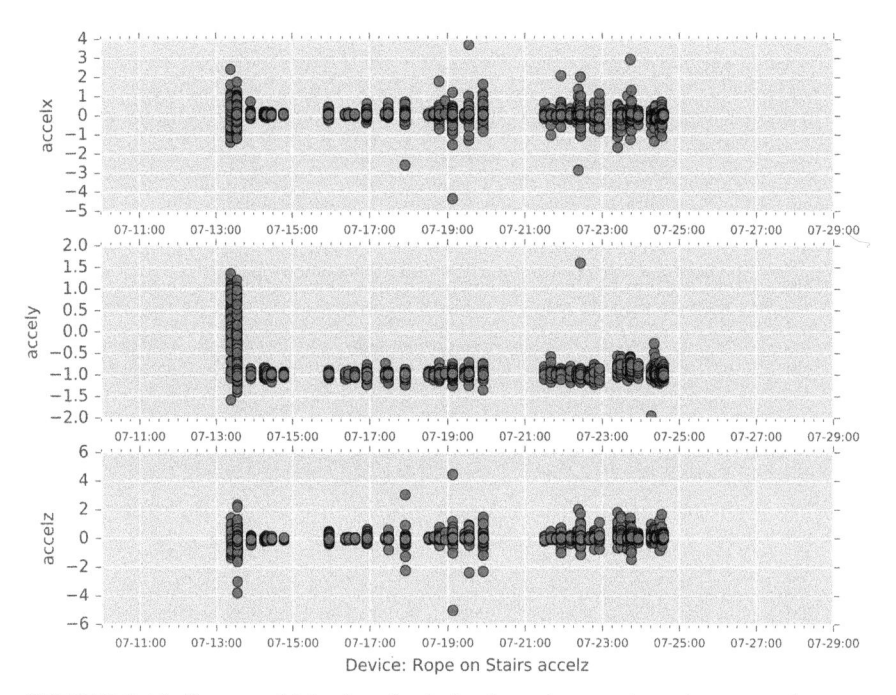

FIGURE 6.10 Resourceful Ageing: Analysis of raw temporal events concerning co-usage of instrumented objects (i.e., relations among nonhuman participants). Data visualization by Yanxia Zhang.

not expect our nonhuman partners to be able to reveal practices of everyday improvisation that are inherently human. Our original hypothesis concerned whether the collected data were able to reveal unusual usage of things, and whether the clusters identified through machine learning analysis were consistent with the

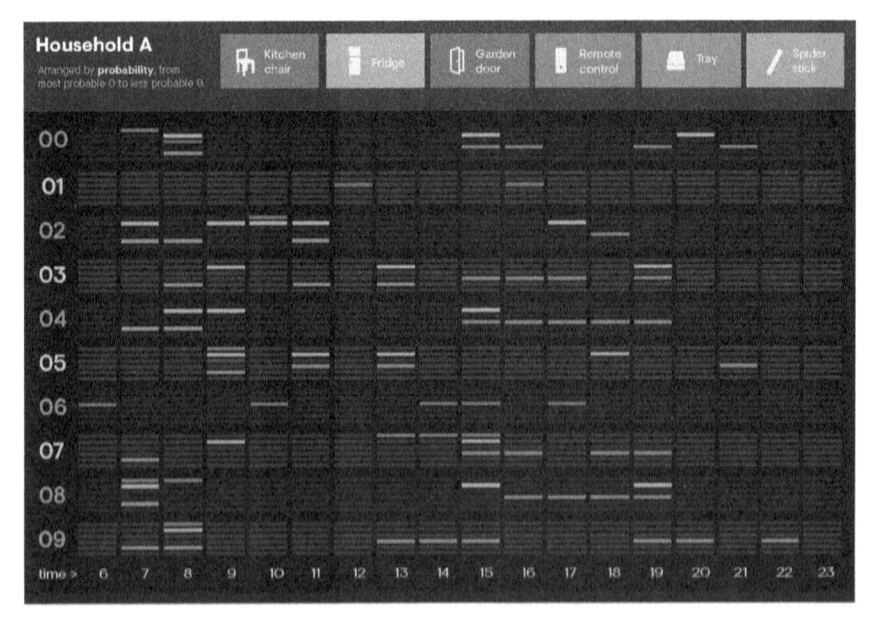

FIGURE 6.11 Resourceful Ageing: Visualization of machine learning interpretation of the co-usage of objects, from high to low probability of occurrence. Data visualization by Philips Design.

strategies observed by humans in the field or actually suggesting new strategies. The expectation was that, through sensors and algorithms, things could give us access to previously unattainable trajectories of use and reveal patterns that could help us ask interesting questions to our human participants.

When nearly a year after data collection we managed to take the patterns generated by machine learning back to our human participants, we indeed obtained new insights about resourcefulness that would have been difficult to obtain otherwise. One of the strategies of resourcefulness that we identified as most prominent in older people was finding your own, unique solutions to challenges: doing things your own way, for your own reasons. These unique solutions deviate from commonly agreed ways of doing. For our participants, they included, for example, eating dinner at their daughter's home or microwaving a meal (both revealed by the absence in the patterns of the fridge around dinner time) or having breakfast in front of the TV with the grandchildren (revealed by the simultaneous use of remote control and fridge in the morning).[19] Identifying these forms of resourcefulness is tricky, because there is no one commonly agreed way of doing that applies in all situations. There is also some form of embarrassment that goes with solutions that participants enact and yet perceive as "uncommon," "strange," or somehow "out of the norm." With this evidence on the table, participants were nudged to reveal a little more information about their everyday lives that might be considered slightly deviant from what is "normal" or expected.

There were also new examples of resourcefulness that came up in the follow-up interviews and not in the data at all. Just talking about resourcefulness—whether triggered by the machine learning patterns or something else—will bring out more examples and knowledge about it. For example, we learned how strategies of resourcefulness—in this case a clever way of getting the coin out of the shopping cart with reduced force in fingers—are eagerly shared among age peers.

More-than-Human Sensemaking and Framing

In this project, the configuring of a more-than-human sensemaking and framing was meant to be the result of a continuous feedback loop between what humans can see and what things can see, where "seeing" here is understood in terms of both what can be observed and how this is interpreted. As speculated in Cila, Giaccardi, Caldwell, Rubens et al. (2015), ethnographic research and machine learning can be complementary. It is difficult for a human ethnographer to see patterns at large scales, whereas a machine (and the computer scientist writing the code) cannot see which patterns are meaningful. This is essentially a question of what inputs matter and why, in a certain situation. We assumed that by looping qualitative data (from human ethnographers in the field) and quantitative data (from thing ethnographers via machine learning), we would have learned something new about how older people use things in everyday life. Unexpected patterns of use would have emerged within the data that was streamed through the interaction between people and things, and things and things, and these would have helped designers identify opportunities for resourcefulness. Though useful insights were eventually generated in the follow-up interviews conducted at the end of the project, the design partnership that configured throughout the project took on a different character.

Confronted with technological limitations and misalignments in the collaborative process,[20] we came to realize that a much more interesting role for our artificial partners in this project was not so much about expanding the processing of the data beyond human capacity and skills and identifying unusual usage patterns within the data. It was instead a more generative role: to suggest probabilities that might constitute openings for different kinds of strategies (and values and norms) to be generated and exchanged. Rather than revealing patterns as "assumed-to-be-real facts" that designers could use for inspiration (like in the first case study), the probabilistic model used for the machine learning analysis was opening up patterns as "possibilities" for objects of design that those taking part in the design process could all contribute to construct, from professional designers and older people to algorithms.

This understanding began to shape in the first phase of the project, when a thing perspective was casted upon the ethnographic fieldwork in elderly homes, in the attempt to identify everyday objects to instrument with sensing capabilities. In this process, driven by human ethnographers but decentered in perspective, we observed that everyday objects become relevant to dispersed practices of resourcefulness when configured in fluid and dynamic arrangements, which change according to the situation of use. These arrangements are constituted not only by spatial proximity (as by positioning in relation to each other, and location within the home) but also by temporal proximity (as in sequences, and when they are used together).[21] We could then observe links between different practices and how resourcefulness is constructed at the overlap of these practices, as objects move across arrangements and become some-thing else. A broom with a piece of tape attached to the stick becomes a handy spider killer, a newspaper moved to a daughter's mailbox becomes a message to communicate well-being, and a metal bar arranged under the bed at night becomes a defense tool to feel safe. Not only materials are reconfigured, but also skills, meanings, and the links between them.

Because taking a thing perspective helps minimize human judgments about what situations may be relevant, memorable, or representative, artificial partners are well suited to reveal misuse, variation, and deviations from norms.[22] Including the perspective of domestic objects in the preliminary ethnographic fieldwork and workshop sessions with older people invited us as humans to explore how an idea of variation could be materialized. We began to conceptualize connected technologies as resources themselves, capable to adapt to changing circumstances in a variety of ways and complement aging competences dynamically. This helped us step away from a focus on the intended use of the technology to be designed and challenged us to explore design as an ongoing process that does not end when the product is released to the market.

These considerations informed the choice of techniques used in the design of the algorithms as well as the way in which machine learning was performed. From both domestic objects and people, we learned during fieldwork and sessions that in order to shift from designing assistive products to designing for resourceful aging, we had to fundamentally step away from solving older people's problems to supporting their improvisational strategies.

For a design partnership to work, machine learning similarly had to step away from its ethnographic role in support of design and embrace the more interventionist and transformative role of future-oriented processes (Smith and Otto 2016). By modeling probabilities and opening up possibilities for new strategies to be experimented, and new values and norms to be established, we realized that machine learning could enable and encourage older people to improvise new strategies. The partnership that this project configured was not so much about identifying design opportunities or alternatives; it was instead about creating possibilities for actualizing always new resources.

Emergent Objects of Design

We decided to pursue these possibilities for empowering older people to be as resourceful with connected technologies as we have observed them to be with their physical domestic objects at home. *Connected Resources* (Figure 6.12) is a series of small, connected devices and an online service that adds digital capabilities to older people's everyday strategies of resourcefulness. Conceptualized as resources (Nicenboim, Giaccardi, and Kuijer 2018), these devices are designed as a family of recombinant sensors and actuators, meant to emulate in physical form and digital functionality the material affordance of the mundane things used by older people in their everyday strategies of resourcefulness. Sensors and actuators can be used alone or together. Once in use, they begin to learn from the way in which they are combined and deployed (Figure 6.13). Via the service (a mobile app) older people

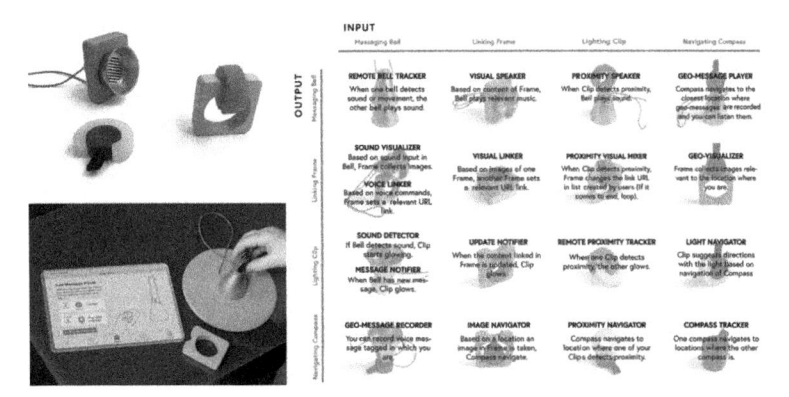

FIGURE 6.12 Resourceful Ageing: *Connected Resources* is a family of sensors and actuators and an online service for adding digital capabilities to older people's everyday strategies of resourcefulness and empowering them in their relation with care technology. Images by Masako Kitazaki.

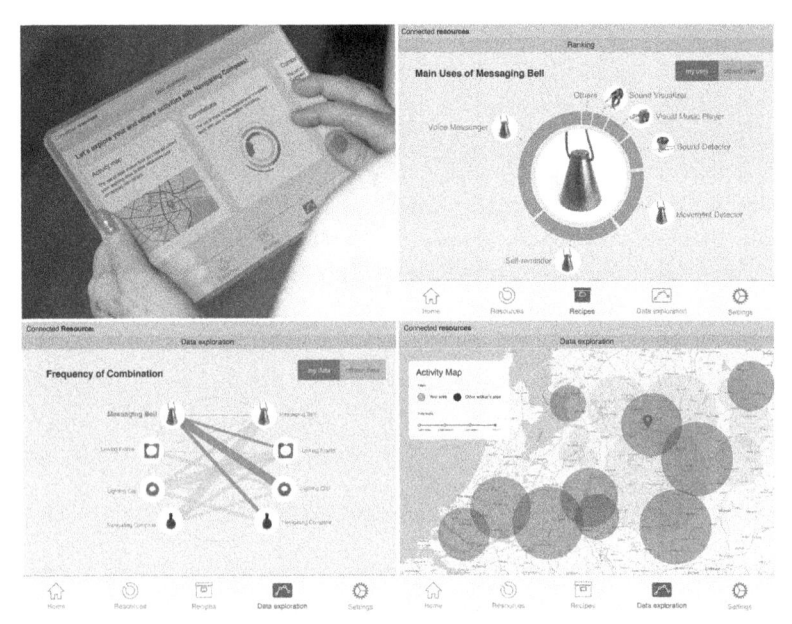

FIGURE 6.13 Resourceful Ageing: Once in use, *Connected Resources* learn from the way in which they are combined and deployed.

122 RELATING TO THINGS

FIGURE 6.14 Resourceful Ageing: Scenario of a resourceful arrangement created by an older woman waiting for a delivery and with a mild hearing impairment, where one object visibly lights up when another remote object detects sound. Movie by Andreas D'Hollandere.

can establish connections between the devices, reflect upon their own strategies, and share their solutions with others. Figure 6.14 is the scenario of a resourceful arrangement created by an older woman waiting for a delivery and with a mild hearing impairment, where one object visibly lights up when another remote object detects sound.

In *Connected Resources*, the design work is not done by the designer alone. Machine learning algorithms are at work too. Possible affordances and performances of the technology are surfaced and arranged into resources, as algorithms work together with older people to empower them in their strategies. Casting things in design meant here for the designer to envision what dimensions of the artifact should stay open, and which instead should be closed so to enable the algorithm to continue the design work at use time.[23] This focus on resourcefulness opened up a design space for interactions between people and things, and all the relations in between (things and things, people and people), which steps away from the prescriptive frameworks of care technology for older people and invites instead creative engagement with both materiality and social norms (in this case, norms concerning what are personally meaningful and socially acceptable ways of complementing one's aging skills; what in the project our human participants often referred to as "normal").

Toward a More-than-Human Design Practice

Experimenting with how we can engage with things, and balance both human and artificial perspectives, is vital to shape future design practices. But what happens when things stop working for us and start working next to us?

As suggested by the cases examined in this chapter, designers can certainly partner with things to expand their capabilities and use "the richness of the artificially broadened context" (Dorst 2015a, 26) to understand the deeper issues that are at play in a situation. Fueled by current developments in the field of artificial intelligence (from increasingly widespread task-specific machine learning algorithms to neural networks), present data-driven design practices emphasize an idea of human augmentation that traces back to Engelbart's 1962 foundational paper "Augmenting Human Intellect." The idea of "a human-machine hybrid built to do more than any person or computer could accomplish alone" is underpinning statements among practitioners of a coming to an end of the era of human-centered design (cf. Milan 2017).[24]

In this chapter, I instead suggest that the partnership between designers and data-enabled, algorithmically powered things should be understood as more than just chasing hard data for making critical decisions in sensitive domains or than working together toward a human-originated, fixed common goal. Fully understanding the assets and benefits of such a partnership requires acknowledging that the uniquely artificial capabilities of things may question our goals by enabling us to access data worlds we have never accessed before, see what we could not see, and call attention to what we thought was marginal or irrelevant. This calls us to stay open to be challenged and surprised. It also requires acknowledging that the design work of contributing a nonhuman point of view does not end with a descriptive account.[25] As suggested programmatically in Giaccardi, Speed, Cila, and Caldwell (2016), *the implications of a thing perspective for design concern fundamentally new alliances for making sense, framing and bringing into existence "things" that do not exist yet—which is at the essence of design work*. It is therefore the hypothetical way of looking at the world in which both humans and nonhumans participate, which configures the scope of design work and generates futures. This was the case when writing social futures for Taiwanese smart mobility or rewriting assistive technology as resource, not aid for our aging future.

As highlighted by the cases, identifying, articulating, and assessing trade-offs represent a unique challenge in pursuing desirable more-than-human design practices. For example, balancing how to set up instrumentation for artificial sensing and analytics in a way that enables you to find out what you did not already know, and yet is carefully crafted to gain access to supposedly relevant data worlds, is a common trade-off we have encountered in our own practice. Trade-offs are the most basic characteristics in design (Simon 1996). As argued in Fischer 2018, trade-offs are often characterized and conceptualized as binary choices. However, exploring the middle ground between these endpoints may help future designers gain a deeper understanding of what balance to strive for when there are no decontextualized sweet spots. Rather than an exclusive focus on human perspectives only, the value of examining trade-offs in more-than-human design practices is grounded in the key objectives this chapter had grappled with. We briefly summarize them here for convenience.

1. Take time and effort to engage with things despite their withdrawing from human perception and understanding

At the beginning of the chapter, I discuss the agency of things and the role they may play in design. At the core of my argument is the idea that things make things too. Enabled by data technologies, things not only perform social practices next to people; they make things. Compared to approaches in design pointing to the ontological gap between things and us, this feminist reconceptualization of performativity as "making" emphasizes engagement over withdrawal. In so doing, it shifts the locus of doing design toward a fundamentally participative relation. This new partnership assumes to spend time with things and painstakingly work together with them to offer different ways of understanding what we know and what we do, humans and nonhumans alike, and ultimately reframe and reconfigure our social and material relations.

2. Balance perspectives informed by capabilities and doings uniquely human and uniquely artificial

I use three case studies to show how balancing uniquely human and uniquely artificial capabilities and doings is of the essence for a more-than-human design partnership. My argument is that this act requires casting things as partners in design, in their being performatively imbricated in how the world actively and continuously configures and reconfigures itself. Ideas of human augmentation or humans and things as independent of each other do not find place in this proposal. As illustrated in the case studies, human and artificial partners have different capabilities and doings (e.g., in unveiling mobility ecosystems, supporting democratized manufacturing, or empowering older people's resourcefulness). These different capabilities and doings enable them to participate in the work doing design with different perspectives, configuring the scope of design work, and embracing the unknown nature of the outcome from different points of view. In a more-than-human design practice, human and artificial partners both participate in sensemaking as well as framing.

3. Account for how things may take on different roles before, during, and after project time

The three different case studies in this chapter also illustrate the spectrum of possible roles things may take in the work of doing design. As implied by the idea of things making things, the central argument here is that the artificial performance of things is not separate from professional design practice. As partners in a more-than-human design practice, things can perform and do design work next to professional designers before, during, and after project time. In the case study about Taiwanese smart mobility, things helped generate insights; they played an explicit role until the designer began to produce design ideas.[26] In the case of the Resourceful Ageing project, things instead both helped generate insights and were conceptualized in the final design as capable of sustaining older people's resourcefulness over time.

4. Problematize the design space in ways that productively enhance, complicate and even challenge what we know and what we do (or how we do it)
Considering things as design partners is different than looking at them as collaborators in achieving human originating purposes. By exerting the ability to access trajectories unattainable to human observation and potentially contesting our worldview, things contribute a different perspective and unique insights that enhance, complicate, and even challenge those of humans. Instead of reinforcing existing blind spots and dominant biases, a thing perspective should instead problematize the design space: unsettle a designer's assumptions, demonstrate the problem to be more uncertain, more nuanced or more complex than originally assumed or regarded. All case studies well illustrate this point, showing for example how social relations and meanings in Taiwanese smart mobility are as important as material and functional factors, or how being resourceful with technology means to older people being independent from technology too.

5. Enable the exploration of practices and objects of design that are not constituted yet but emerge as appropriate and desirable
Considering things as partners can help us see what is not immediately apparent or may fall outside of our sense of relevance. By problematizing and potentially contesting what we take for granted, a thing perspective opens up a perspective on the emergent that goes beyond the descriptive and emerges at the intersection of the trajectories that things give access to and the analysis that humans bring to it. In a more-than-human design practice, the aim of a thing perspective is fundamentally to enable the exploration of practices and objects of design that are not constituted yet but emerge as appropriate and desirable in response to more-than-human sensemaking and framing.

Conclusion

In a world in which we come to live with things, and things with us in ways that blur distinctions between producer and produced, subject and object, us and them, design must go beyond a narrow focus on humans.

Things become, as does our knowledge of them. It follows that our primary focus should not be on the ontologies of things but on their ontogenies, not on philosophies but on generations of being. This shift of focus has important political ramifications. For it suggests that things are far from closed to one another, each wrapped up in its own, ultimately impenetrable world of being. On the contrary, they are fundamentally open, and all are participants in one indivisible world of becoming. Multiple ontologies signify multiple worlds, but multiple ontogenies signify one world. And since, in their growth or movement, the things of this world answer to one another, or correspond, they are also responsible. All responsibility depends on responsiveness. (Ingold 2017)

Casting things as partners in design, bringing their artificial capabilities and nonhuman perspectives to bear on how problems are framed and addressed, shifts the emphasis in design research concerned with data technologies from the functionality of the designed artifact to the intentionality of design work and its trade-offs. This more-than-human turn offers designers an avenue to reshape human-technology relations in the widening world of design practice. Intrinsically vibrant and transformative, the more-than-human design practice proposed and experimented here is fundamentally defined by the characteristics of its process and the responsiveness of those engaged in the process.

Acknowledgments

Writing this chapter would not have been possible without the friendship and support of Chris Speed and Johan Redström, as well as the intellectual contributions and design work of many colleagues and students at the Connected Everyday Lab, in particular Iohanna Nicenboim, Nazli Cila, and Lenneke Kuijer. This research was supported by TU Delft Technology Fellowship (2012–17), STW Research through Design (2015/16734/STW), and NTU IoX Center.

Notes

1. Anne-Marie Willis (2006) has made a strong argument that things also make humans in her work on ontological designing, and similarly Tony Fry (2012). This perspective, however, is not typical of "ordinary" designers.
2. Heidegger (1967).
3. Latour and Weibel (2005).
4. A connected health device, for example, is not only a product service that helps people track and monitor their diet. It is also part of broader processes of preventive care, and it may find itself in a new industry, such as the insurance industry, connected horizontally to products that could never have been connected before (Neese 2015).
5. To clarify the philosophical distinction between post phenomenology and object-oriented ontology on matters of "withdrawal," it is important to note that

in post-phenomenology, the idea of withdrawal is understood "as only one of many potential ways that technological mediation shapes the contours of a user's overall experience" (Rosenberger and Verbeek 2015, 23). Post phenomenologists also investigate "what stands forward in addition to what withdraws, what demands attention, what remains on the fringes" (ibid.) within a given human-technology relation. On the contrary, object-oriented ontology assumes things (more precisely "objects") to exist independently of human experience, and ontologically not exhausted by their relations with humans or other objects (Harman 2002).

6 Co-performance as elaborated in Kuijer (2018) confines artificial agency at use time.
7 As argued by James J. Gibson (1979), mobility is one of the cornerstones of humanity. From crawling to walking, from the saddle to the hover board, means of movement and transport change the ways we live and the relationship among people, things, and environments.
8 The Taiwan-based company Gogoro's Smartscooter™, for example, uses over eighty sensors to continuously learn people's riding patterns and suggest customized ways to save energy.
9 *Interview with Scooters* is a graduation project by Wen-wei Chang conducted at National Taiwan University of Science and Technology under the joint supervision of the author, Lin-Lin Chen and Rung-Huei Liang (cf. Chang 2016).
10 For additional examples and further insights on the dynamic, situated nature of the relationship between Taiwanese scooters and scooterists, and associated meanings, see Chang et al. (2016).
11 *Stimulating Creative Dialogues Between Humans and Things* is a graduation project by Tal Amram conducted at Delft University of Technology under the supervision of the author and Jan Willem Hoftijzer (cf. Amram 2016).
12 For details about the different studies and a complete report of their findings, see Amram (2016).
13 *Resourceful Ageing* is a project funded by STW under the Research Through Design program (2015/16734/STW), http://www.resourcefulageing.nl/. The project ran from June 2016 to June 2018, and it involved four senior researchers, three postdocs, and a range of technical assistants and master students.
14 Project partners included Delft University of Technology (coordination), Eindhoven University of Technology, Avans University of Applied Sciences, and Philips.
15 This list is nonexhaustive. For a complete list, cf. Hung and Zhang (2018).
16 For more details about the design research techniques used in this phase of the project, cf. Nicenboim et al. (2018).
17 For technical details about sensor data collection and machine learning analysis, cf. Hung and Zhang (2018).
18 For additional examples of resourcefulness captured through ethnographic fieldwork, cf. Giaccardi and Nicenboim (2018).
19 For more details about the results of the "closing-the-loop" interviews, cf. Giaccardi and Nicenboim (2018).
20 For a discussion about the limitations and challenges of interdisciplinary research projects in the space of data-enabled design, cf. Giaccardi and Nicenboim (2018).
21 For an in-depth discussion of spatial and temporal arrangements in elderly homes and additional examples, cf. Giaccardi and Nicenboim 2018.
22 Please be noted that misuse is used here provocatively, from the perspective of the professional designer's original intention (Brandes, Stich, and Wender 2008). At use time, there are no misuses, only variety of use (Hui 2017).

23 For more details on the dimensions of openness and closure in the design of Connected Resources, and how these are necessary to support both "variety of use" and "variety in use," refer to Kitazaki (2018).
24 In the field of design for healthcare and social well-being, where entire populations are targeted, the use of AI is seen as going hand in hand with the chasing of hard data for making critical, evidence-based design decisions.
25 Contemporary anthropological approaches engaged with design work clearly posit that the potential of anthropology is not in presenting a solution to a design context, as not all problems have simple answers (Dourish 2006). Greater impact is achieved in shaping the way that a phenomenon is understood in the design process, with those involved in the design process. The field of design anthropology brings this further and concerns itself with collaborative future making, with a strong commitment to intervention and transformation of social realities (Smith and Otto 2016). Our work in thing-centered design approaches has always been aligned with these positions.
26 We could argue that the designed accessories for smart scooters do turn the scooter into a thing capable to continue playing a role in design work, for example, by enabling the scooter to tune its performance in order to shape a distinct relation to users. But the designer in his conceptualization of a more-than-human design practice did not explicitly intend this.

References

Amram, T. 2016. *Stimulating Creative Dialogues between Humans & Things*. Delft University of Technology.

Appadurai, A. 1986. *The Social Life of Things: Commodities in Cultural Perspective*. Edited by A. Appadurai. Cambridge; New York: Cambridge University Press.

Atkinson, P. 2006. "Do It Yourself: Democracy and Design." *Journal of Design History* 19 (1): 1–10.

Barad, K. 2003. "Posthumanist Performativity: Toward an Understanding of How Matter Comes to Matter." *Signs: Journal of Women in Culture and Society* 28 (3): 801–831. http://doi.org/10.1086/345321.

Bennett, T. and P. Joyce 2013. *Material Powers: Cultural Studies, History and the Material Turn*. London and New York: Routledge. http://doi.org/10.4324/9780203883877.

Bogost, I. 2012. *Alien Phenomenology, or What It's Like to Be a Thing*. Minneapolis, MN: University of Minnesota Press. http://doi.org/10.1017/CBO9781107415324.004.

Brandes, U., S. Stich, and M. Wender 2008. *Design by Use: The Everyday Metamorphosis of Things*. Birkhauser.

Brown, B. 2001. "Thing Theory." *Critical Inquiry* 28 (1): 1–22. http://doi.org/10.1086/449030.

Bryant, L. 2011. *The Democracy of Objects*. Open Humanities Press.

Busbea, L. 2009. "Metadesign: Object and Environment in France, c. 1970." *Design Issues* 25 (4): 103–119.

Chang, W.-W. 2016. *Interview with Scooters: A "First-Thing Perspective" Thing Ethnography Research and Design of the Scooters in Taiwan*. National Taiwan University of Science and Technology.

Chang, W.-W., E. Giaccardi, L.-L. Chen, and Liang Rung-Huei. 2016. "'Interview with Things': A First-Thing Perspective to Understand the Scooter's Everyday Socio-Material Network in Taiwan." In *DIS 201—Proceedings of the 2016 ACM Conference on Designing Interactive Systems: Fuse*, 1001–1012. New York: ACM Press.

Cila, N., E. Giaccardi, M. Caldwell, N. Rubens, F. Tynan-O'Mahony, and C. Speed. 2015. "Listening to an Everyday Kettle: How Can the Data Objects Collect Be Useful for Design Research?" In *Reframing Design—Proceedings of the 4th Participatory Innovation Conference (PIN-C 2015)*, 500–506. The Hague University of Applied Sciences.

Cross, N. 1981. "The Coming of Post-industrial Age." *Design Studies* 2 (1): 3–7.

De Michelis, G., G. Jacucci, I. Wagner, P. Ehn, P. Linde, and T. Binder. 2011. *Design Things*. Cambridge, MA: MIT Press.

Di Salvo, C. and J. Lukens. 2011. "Nonanthropocentrism and the Nonhuman in Design: Possibilities for Designing New Forms of Engagement with and through Technology." In M. Foth, L. Forlano, C. Satchell, and M. Gibbs, eds., *From Social Butterfly to Engaged Citizen*, 440–460. Cambridge, MA: MIT Press.

Dorst, K. 2015a. "Frame Creation and Design in the Expanded Field." *She Ji* 1 (1): 22–33.

Dourish, P. 2006. "Implications for design." In *CHI '06 Proceedings of the SIGCHI Conference on Human Factors in Computing Systems*, 541–550. New York: ACM Press.

Ehn, P. 2008. "Participation in Design Things." In *PDC '08 Proceedings of the Tenth Anniversary Conference on Participatory Design 2008*, 92–101. New York: ACM Press. http://doi.org/10.1145/1795234.1795248.

Engeström, Y. and F. Blackler. 2005. "On the Life of the Object." *Organization* 12 (3): 307–330.

Fischer, G. 2018. "Exploring Design Trade-Offs for Quality of Life in Human-Centered Design." *ACM Interactions* 25 (1): 26–33.

Fry, T. 2012. *Becoming Human by Design*. London: Bloomsbury.

Gatt, C. and T. Ingold. 2013. "From Description to Correspondence: Anthropology in Real Time." In *Design Anthropology: Theory and Practice*, edited by W. Gunn, T. Otto, and R. C. Smith, 139–158. London: Bloomsbury Academic. http://doi.org/978-0-8578-5369-1.

Giaccardi, E. 2005. "Metadesign as an Emergent Culture of Design." *Leonardo* 38 (4): 342–349.

Giaccardi, E. 2019. "Histories and Futures of Research through Design: From Prototypes to Connected Things." *International Journal of Design*, 13 (3): 139–155.

Giaccardi, E. and I. Nicenboim 2018. *Resourceful Ageing: Empowering Older People to Age Resourcefully with the Internet of Things*. Delft University of Technology, Delft, The Netherlands.

Giaccardi, E., C. Speed, and M. Netten. 2016. *Things2Things: Designing in the Connected Everyday*. Delft University of Technology, Delft, The Netherlands.

Giaccardi, E., L. Kuijer, and L. Neven. 2016. "Design for Resourceful Ageing: Intervening in the Ethics of Gerontechnology." In *Proceedings of DRS 2016, Design Research Society 50th Anniversary Conference*, June 27–30, Brighton, UK.

Giaccardi, E., C. Speed, J. Grossen, and P. van Allen. 2014. "The Things About Design: of Ghosts, Spirits and Material Practices." In *Companion to DRS 2014: Design's Big Debates*, edited by J. Redström, E. Stolterman, A. Valtonen, and H. Wiltse, 175. Umeå: Umeå University.

Giaccardi, E., C. Speed, N. Cila, and M. Caldwell. 2016. "Things as Co-Ethnographers: Implications of a Thing Perspective for Design and Anthropology." In *Design Anthropological Futures*, edited by R.C. Smith, K.T. Vangkilde, M.G. Kjaersgaard, T. Otto, J. Halse, and T. Binder, 235–248. London: Bloomsbury Academic.

Giaccardi, E., N. Cila, C. Speed, and M. Caldwell. 2016. "Thing Ethnography: Doing Design Research with Non-Humans." In *DIS '16 Proceedings of the 2016 ACM Conference on Designing Interactive Systems*, 377–387.

Gibson, J.J. 1979. *The Ecological Approach to Visual Perception*. Boston, MA: Houghton Mifflin.

Grudin, J. 2017. *From Tool to Partner: The Evolution of Human-Computer Interaction*. Morgan & Claypool.

Gunn, W. and J. Donovan. 2012. "Design Anthropology: An Introduction." In *Design and Anthropology*, edited by W. Gunn and J. Donovan. London and New York: Routledge.

Harman, G. 2002. *Tool-Being: Heidegger and the Metaphysics of Objects*. Peru, IL: Open Court.

Harman, G. 2009. *Prince of Networks: Bruno Latour and Metaphysics*. Prahran, Victoria [Australia]: re.press.

Hebdige, D. 2001. *Object as Image: The Italian Scooter Cycle*. London and New York: Routledge.

Hodder, I. 2012. *Entangled: An Archaeology of the Relationships Between Humans and Things. Entangled: An Archaeology of the Relationships Between Humans and Things*. Hoboken, NJ: Wiley-Blackwell. http://doi.org/10.1002/9781118241912.

Hui, A. 2017. "Variation and the Intersection of Practices." In *The Nexus of Practices : Connections, Constellations, Practitioners*, edited by A. Hui, T. Schatzki, and E. Shove, 52–67. London and New York: Routledge.

Hung, H. and Y. Zhang. 2018. "Using Topic Models to Mine Everyday Object Usage Routines Through Connected IoT sensors." *E-Print*.

Hunt, J. 2005. "A Manifesto for Postindustrial Design." *I.D. Magazine*, December.

Ihde, D. 1993. *Philosophy of Technology: An Introduction*. New York: Paragon House Publishers.

Ingold, T. 2017. "On Human Correspondence." *Journal of the Royal Anthropological Institute* 23 (1): 9–27.

Iqbal, M. 2018. *Thinking in Services: Understanding and Exploring the Expanding Universe of Services*. BIS Publishers.

Kaptelinin, V. and B.A. Nardi. 2006. *Acting with Technology: Activity Theory and Interaction Design*. Cambridge, MA: MIT Press. http://doi.org/10.1897/IEAM_2009-036.1.

Kitazaki, M. 2018. *Connected Resources: A Research Through Design Approach to Designing for Older People's Resourcefulness*. Delft University of Technology.

Kolko, J. 2010. "Sensemaking and Framing: A Theoretical Reflection on Perspective in Design Synthesis." In *Design Research Society 2010*.

Kuijer, L. 2018. "Automated Artefacts as Co-Performers of Social Practices: Washing Machines, Laundering and Design." In *Social Practices and Dynamic Non-Humans: Nature, Materials and Technologies*, edited by C. Maller and Y. Strengers, 193–214.

Kuijer, L. and E. Giaccardi. 2018. "Co-performance: Conceptualizing the Role of Artificial Agency in the Design of Everyday Life." In *CHI '18 Proceedings of the 36th Annual ACM Conference on Human Factors in Computing Systems*. New York: ACM Press.

Kuijer, L., I. Nicenboim, and E. Giaccardi. 2017. Conceptualising Resourcefulness as a Dispersed Practice. In *Proceedings of the 2017 Conference on Designing Interactive Systems (DIS '17)*, 15–27. New York: ACM Press. http://doi.org/http://dx.doi.org/10.1145/3064663.3064698.

Kuznetsov, S. and E. Paulos. 2010. "Rise of the Expert Amateur: DIY Projects, Communities, and Cultures." In *Proceedings of the 6th Nordic Conference on Human-Computer Interaction: Extending Boundaries*. New York: ACM Press.

Lai, Y. 2010. *Riding Under the State Gaze: The Historical Analysis of Taiwan Autobike from the Japanese Colonial Period to 1970*.

Lin, S. 1998. "Taiwan Motorcycle History." *Taiwan Motorcycle Research and Development Security Promotion Association*.

Marenko, B. 2014. "Neo-Animism and Design: A New Paradigm in Object Theory." *Design and Culture* 6 (2): 219–241. http://doi.org/10.2752/175470814X14031924627185.

McVeigh-Schultz, J., J. Stein, J. Watson, and S. Fisher. 2012. "Extending the Lifelog to Non-Human Subjects: Ambient Storytelling for Human-Object relationships." *Proceedings of the 20th ACM International Conference on Multimedia (MM '12)*, 1205–1208. http://doi.org/10.1145/2393347.2396419.

Milan, M. 2017. "The Next User You Design for Won't Be a human." *FastCompany*. November 20, 2017. https://www.fastcompany.com/90146967/the-next-user-you-design-for-wont-be-a-human.

Miller, D. 2009. *The Comfort of Things*. Cambridge: Polity Press.

Neese, M. 2015. "What Is a Product? How a New Definition Is Leading Us Toward a Place-Based Design process." In *EPIC Perspectives*. July 27, 2015. https://www.epicpeople.org/what-is-a-product/.

Nicenboim, I., E. Giaccardi, and L. Kuijer. 2018. "Designing Connected Resources for Older People." In *Proceedings of DIS' 18*, 413–425. New York: ACM Press.

Piaget, J. 1959. *The Language and Thought of the Child*. London: Routledge & Kegan Paul.

Redström, J. and H. Wiltse. 2018. *Changing Things: The Future of Objects in a Digital World*. London and Oxford: Bloomsbury Academic.

Rosenberger, R. and P.-P. Verbeek. 2015. "A Field Guide to Postphenomenology." In *Postphenomenological Investigations: Essays on Human-Technology Relations*, edited by R. Rosenberger and P.-P. Verbeek, 9–41. Lexington Books.

Rozendaal, M. 2016. "Objects with Intent: A New Paradigm for Interaction Design." *Interactions* 23 (3): 62–65.

Simon, H.A. 1996. *The Sciences of the Artificial*. Cambridge, MA: MIT Press.

Smith, R.C. and T. Otto. 2016. "Cultures of the Future: Emergence and Intervention in Design Anthropology." In R.C. Smith, K.T. Vangkilde, M.G. Kjaersgaard, T. Otto, J. Halse, and T. Binder, *Design Anthropological Futures*. Bloomsbury Academic.

Tonkinwise, C. 2005. "Is Design Finished? : Dematerialisation and Changing Things." *Design Philosophy Papers* 3 (2): 99–117. http://doi.org/10.2752/144871305X13966254124437.

Tonkinwise, C. 2015. "What Things to Teach Designers in Post-Industrial Times?" *EPIC Perspectives*, August 25, 2015. https://www.epicpeople.org/what-things-to-teach-designers-in-post-industrial-times/.

Turkle, S. 2007. *Evocative Objects: Things We Think With*. (S. Turkle, Ed.). MIT Press. http://www.amazon.com/dp/0262201682.

Verbeek, P.-P. 2005. *What Things Do: Philosophical Reflections on Technology, Agency, and Design*. University Park, PA: Penn State University Press. http://doi.org/10.1017/CBO9781107415324.004.

Willis, A.-M. 2006. "Ontological Designing." *Design Philosophy Papers* 4 (2): 69–92.

PART THREE

CONTROLLING THINGS THAT CONTROL US

It is imperative that we find ways to conceive of how technological mediation at times constitutes a form of control. One of the ways that we relate to the world through our technologies is by using them to control others. One of the ways that the objects of our world relate back to us is by the manners in which we are controlled by them.

In this part, we begin to think about the ways that technological relations can be relations of influence, relations of surveillance, relations of governance, or relations of force. In this way, our relations to the things that relate back to us can raise social, ethical, and political concerns. How should we think about the ways that technologies are used by some people to influence the behaviors of others? How should we conceive of the dynamics in which our technologies make us do things, sometimes without our knowledge, and sometimes against our will? How should we reckon with technologies that take part in agendas that exclude, discriminate, or victimize? How should we evaluate technologies that, by their very design, leave us with little choice in how we may act?

It is important as well to consider the potential for technologies to play a part in resisting systems of unjust control. In this exploration of our relations to the things that relate back to us, we must attempt to conceive of the ways that technologies have the potential to open up new possibilities for action in the face of agendas that look to restrict everything except their prescribed behaviors. We must find ways to think about the potential for technological mediation to constitute forms of critique, subversion, and resistance.

7 HOSTILE DESIGN AND THE MATERIALITY OF SURVEILLANCE

Robert Rosenberger

There has been a recent upsurge in interest, especially in online journalism, in what could be called "hostile design." Roughly, this refers to the design of objects with the aim of discouraging specific uses of public space, frequently with the goal of pushing a particular population out of public space entirely. Hostile designs often function in accord with rules and laws that target these same populations and take part in larger social and political agendas. Some examples include things like spikes set into ledges to deter homeless people from sitting there or benches redesigned in a way that prevents one from lying down and sleeping across them. Indeed, the homeless, or the "unhoused" as some prefer, are the primary population targeted by hostile design, but far from the only one.

Security cameras are also regularly listed as a paradigmatic example of hostile design. But they are a strange fit. Cameras do not bump into or otherwise physically obstruct people the same way as does, say, a spike on a ledge or dividers on a bench. In what follows in this chapter, I consider what it means to conceive of surveillance technology as a form of hostile design. The specific form of surveillance I consider here is security cameras in public spaces. While much work has been done on the topics of surveillance generally in its many forms, I'll focus on public-space cameras as concrete material technologies.

Across a series of works, I have been developing an account of hostile design. In particular, I have drawn together ideas from the philosophy of technology and social theory to conceive of the various ways instances of hostile design shape user experience, and I have developed a critique of anti-homeless design in particular (e.g., Rosenberger 2014; Rosenberger 2017a, 2017b, 2018). I argue that it is helpful to think of hostile designs as mediators of experience, objects that are open to multiple uses and that have been redesigned for the purpose of closing off particular usages in accord with particular agendas that can be drawn out and scrutinized with tools from political and social theory. To do so, I use ideas from phenomenological philosophy, especially the "postphenomenological" school of thought. Ideas from postphenomenology are useful for describing the details of the experience of human-technology relations and useful as well for conceiving of how technologies can mediate experience in multiple ways. When combined with insights from social theory, and in particular the "script theory" perspective from the field of science and technology studies, these ideas can be assembled into the beginning of an account of hostile design.

The nascent discussion over what I'm referring to here as "hostile design" builds on the work of critics that have sought to reveal the politics of public spaces, as well as a history of thought and critique on "defensive spaces" and "broken windows" policing schemes (e.g., Newman 1973; Foucault 1977; Davis 1990; Flusty 1994; Low 2003; Minton 2012; Mitchell 2014). While raising important issues, the contemporary discussion remains disjointed and still emerging (Rosenberger forthcoming). This is reflected in the range of terminology on offer in attempt to capture these phenomena, including "hostile architecture," "disciplinary architecture," "unpleasant design," "architectural exclusion," "cruel design," and "defensive design," among others (e.g., Léopold 2013; Savicic and Savic 2013; Quinn 2014; Schindler 2015; Chellew 2016; Petty 2016; Rosenberger 2017a).

As mentioned, one paradigmatic example is what could be called "anti-sleep benches," that is, benches to which have been added seat dividers or some other design feature that functions to prevent people from lying across and sleeping on them (e.g, Figure 7.1). Another paradigmatic example of hostile design is what

FIGURE 7.1 Subway bench, New York City, United States (photo by author).

FIGURE 7.2 Ledge spikes, San Francisco, United States (photo by author).

could be called "anti-homeless spikes." These are spikes added to ledges and other surfaces to dissuade people from resting there (e.g., Figure 7.2).

There are also what are called "skatestoppers." Skateboarding tricks often involve sliding or "grinding" your board across a curb, ledge, handrail, or some other surface. Skatestoppers are small metal nubs which are attached to these surfaces to discourage skateboarders from riding in the area, and a small skatestopper manufacturing industry has emerged (Rosenberger 2018). This example draws out the inherent value dimension of hostile design. Skateboarders are a very different target population than the unhoused. We can imagine someone who acknowledges that skateboarders are sometimes targeted by hostile design and yet is less (or perhaps more) critical of anti-skateboarding design than they are of anti-homeless design. That is, not all types of hostile design, nor all individual instances of hostile design, must be evaluated the same way.

There are many other examples of hostile design. These can include trashcan lids to prevent picking, fire hydrant locks that deter unauthorized water access, noise machines that deter loitering, fences that close off areas, among many others. As mentioned, it is important to remember that any individual instance of hostile design must be understood in terms of its potential role in a larger agenda, an agenda which might include a pattern of other designs, as well as corresponding laws and social conventions. It is important to remember too that hostile designs are not always the last word on an issue. Resistance efforts may arise, from art installations that call attention to pervasive anti-homeless measures, to the development of techniques for picking hydrant locks, to the vandalism of skatestoppers with crowbars.

This brings us to a central question of this chapter: What should it mean to understand public-space security cameras as a form of hostile design? Although they are routinely listed as one of the main examples of these kinds of control-though-design measures, we will see that they enact their hostility in a very different manner than most of the other examples reviewed. But in order to draw out these issues for the analysis of security cameras, let us first consider how the "postphenomenological" and "script theoretical" frameworks can be used to develop an account of hostile design.

Inscribing Stabilities

A central project of work in the fields of science and technology studies (STS) and the philosophy of technology is the attempt to understand the various ways that technologies are shaped by cultural forces, and how technologies in turn shape people's lives. The main concept I will use here for approaching these kinds of technological dynamics is the notion of "multistability." This idea, brought to us by the postphenomenological philosophical perspective, is the claim that technologies can always maintain multiple relationships with users, fitting differently into different contexts.

The postphenomenological school of thought, founded by American philosopher Don Ihde, offers tools for thinking about human-technology relations (e.g., Ihde 2009; Verbeek 2011; Rosenberger and Verbeek 2015; Wellner 2016; Rosenberger 2017a; Van Den Eede et al. 2017). This perspective brings together insights from phenomenological and pragmatist philosophy to provide an account of how human experience is shaped by technology usage. From phenomenology, this perspective draws a concrete focus on human bodily perceptual experience. From pragmatism, it draws ontological and epistemological commitments to anti-essentialism and nonfoundationalism. In the view of postphenomenology, technologies play a "mediating" role in human experience, transforming the ways a user can perceive and act on the world. A hammer transforms a user's ability to strike nails. Eyeglasses transform a wearer's ability to see. A bench provides a place to sit.

Postphenomenology's conception of "multistability" accords with its pragmatic commitment to anti-essentialism. Under this view, any technology should be understood to be multistable, that is, open to different uses in different contexts and open to being meaningful in different ways to different users. Technologies do not have a fixed essence that somehow determines what single purpose they must serve, what single meaning they must reveal, or what exclusive effects they must have on our lives. Ihde writes that "technologies in use appear differently to beginners compared to skilled users. And as one might expect, multiple outcomes also are likely; technologies tend to be *multistable*" (2009, 44). Case studies on the multistability of technology have included everything from emotionally assistive robots, to the history of archery, to pap smear cancer screening samples, to the evolution of cell phone design, to the experience of "frozen" computer screens, to organ donation protocols (Ihde 2009; Rosenberger 2009; Forss 2012; Hasse 2015; Rosenfeld 2015; Wellner 2016).

I suggest that explicitly conceiving of public-space technologies as multistable can help call attention to some of their alternative stabilities. The bench is not only a place to sit; it can be used as a bed. The garbage can is not only a receptacle; it can be used as a potential resource for discarded food or for recyclable materials that can be turned in for cash. However, this gets us only partway to an account of hostile design. To capture the dynamics of hostile modifications to technologies in public spaces, we need to integrate these ideas with social theory.

Madeleine Akrich's "script theory" provides an ideal starting point. What we need here is not only a conception of technology's capacity to play multiple roles for different people but also a conception of how technologies may function as a part of a larger social agenda. Akrich provides the helpful understanding of technology as following a kind of social "script." As she explains, "A large part of the work of innovators is that of '*inscribing*' this vision of (or prediction about) the world in the technical content of the new object" (1992, 208). When a technology is modified to better enact the purposes of a social agenda, it can be said to be inscribed to better follow its role in a social script. In this way, technologies—or "nonhumans" as they are also called within actor-network theory (a major STS account with which script theory is often associated)—can be redesigned to align with human interests, and together form social networks. Akrich writes, "Technical objects participate in building heterogeneous networks that bring together actants of all types and sizes, whether human or nonhuman" (1992, 206).[1]

My suggestion is that we can develop a basic account of hostile design by combining these postphenomenological insights into technological multistability and these script theoretical insights into social inscription. If we understand technologies to always be open to multiple stable relationships with users, then we can ask under what social and material conditions could users be discouraged from taking up a particular stability that would be available otherwise. As we see in the various examples of hostile design reviewed above, particular uses of public-space objects are sometimes curtailed through strategic redesigns of those objects. A handrail may be open to both the stability of its designed usage and an alternative stability of providing a place to grind one's skateboard. Skatestoppers are added to handrails to shut down that latter stability. A normal bench may be open to both its designed usage for providing a place to sit and an alternative stability in which one uses it as a place to sleep. Seat dividers are added to benches to foreclose that sleeping option. My suggestion is that we can understand these strategic redesigns as material inscriptions.

These inscriptions thus enlist the public-space device into a particular network, a network whose agenda includes targeting a particular minority group and discouraging a particular usage of that device. Insofar as the script followed by the inscribed device can be considered a hostile one, the inscription is an example of hostile design. Thus under this account, hostile designs can be understood as multistable public-space objects that have undergone material inscription for the purpose of "closing off" a particular stability in accord with a network's hostile agenda.[2] And yet, activist resistance efforts always remain possible, with attempts to "reopen" stabilities that had been closed off by a network enacting a hostile agenda.

It should be noted that these analyses shift when we change the focal point of our investigation. We can zoom in on the inscriptions themselves and consider their own multistability and their potential for having multiple effects at the same time. For example, in addition to having an allegedly hostile effect of deterring trash picking, garbage can lids can additionally function to keep out rainwater or deter animals from entering. An armrest added to a bench of course provides a place to rest one's arm. An armrest also can aid the disabled or the elderly in sitting down. It can serve as a divider to the next seat. But an armrest can also have a hostile effect of cutting off the bench as a place to sleep, a point occasionally noted by retail websites and best practice bus stop design guidelines (Rosenberger 2017a). We can zoom out as well to consider how multiple instances of hostile design, combined with local policies, can function together to deter targeted groups from making use of entire areas.

Are Security Cameras Hostile?

While work on hostile design in general remains disparate, issues of surveillance in particular are a major topic of academic investigation. The field of surveillance studies explores the variety of ways that surveillance practices shape different parts of society (e.g., Lyon 1994; Lyon 2001; Levin et al. 2002; Friesen et al. 2009; Monahan 2010; Ball et al. 2012; Marx 2015). Work in this field addresses a vast set of issues, including everything from the relation of surveillance to nationhood, security, and privacy. In addition to public space, surveillance studies scholars investigate an array of contexts, including the workplace, schools, consumer settings, online communications, and targeted advertising. Issues of surveillance are at work in everything from international espionage, to police body cams, to drones, to smartphones, to the census.

Here I restrict focus to only public-space cameras, also sometimes called CCTV (Closed-Circuit Television), and in particular how their specific materiality and socially established meanings mediate the experience of the surveilled (e.g., Figures 7.3 and 7.4). While this is only a small subset of the work of surveillance studies, there is a bustling line of research on which to build (e.g., Norris and Armstrong 1999; Koskela 2000; Goold 2004; von Silva-Tarouca Larsen 2011;

FIGURE 7.3 Tube platform security camera, London, England (photo by author).

FIGURE 7.4 Outdoor security camera mounted on a pole and protected by pointed fencing, Winchester, England (photo by author).

HOSTILE DESIGN AND THE MATERIALITY OF SURVEILLANCE

Goold et al. 2013). In addition, I restrict focus here to the question of how public-space cameras can at times constitute instances of hostile design.

While the usage of public-space cameras of course does not always reduce to the targeting of one particular group, such cameras are indeed at times enrolled into hostile agendas. As Clive Norris and Gary Armstrong note in their critical history of CCTV in Britain, surveillance systems have frequently been used to monitor those who appear not to belong in the surveilled space based in part on

> a normative ecology of place which singled out certain people and behaviours as inappropriate. This was found to be less influenced by strictly crime-related concerns than the commercial image of city centre streets which saw certain people being defined as "other." Thus drunks, beggars, the homeless, street traders were all subject to intensive targeted surveillance. (1999, 197–198)

In a review of disciplinary practices enacted on the unhoused population across Europe, Joe Doherty and colleagues identify a clear pattern. They report: "The surveillance, on the streets and in shelters, of those who are homeless is a distinctive feature of the contemporary city; homeless people are today among the most surveyed and scrutinised of marginal groups" (Doherty et al. 2008, 307). Just as the unhoused are a primary target of hostile design in general, they are correspondingly one for surveillance technologies in particular.

As noted above, public-space security cameras are routinely listed as paradigmatic examples of hostile design. But cameras are in some ways an outlier; they don't physically bump against a targeted person, but instead merely watch. How should we understand the form of the hostility enacted by security cameras? When cameras participate in hostile agendas, by what mechanism is this hostility leveled? What is the "logic" of the hostility of public-space security cameras?

Most (although not all) examples of hostile design can be understood as a form of imposition. That is, many of the examples of hostile design we have considered above function by somehow impeding a particular usage of an object. Many of these examples operate by what we could call a *logic of "physical imposition,"* materially shutting down or interrupting a user's bodily engagement with objects and space in a particular manner. The skatestoppers get in the way of grinding. The armrests or seat dividers attached to a bench present an impediment to lying down. The hydrant locks sit securely in the path of anyone attempting to access the water without exclusively available tools. Other examples we've reviewed instead operate through what could be called a *logic of "sensory imposition,"* presenting an annoying or otherwise unpleasant sensation, often of the auditory variety. The noise machines audible only to the young operate by this logic, obstructing youthful loiterers from comfortably occupying the space in range. A loud noise broadcast throughout the night in a park presents a sensory imposition to anyone with normal hearing, deterring them from sleeping there.

But public-space cameras do not fit this pattern in which the hostility involves a form of physical or sensory imposition enacted by the devices themselves. When cameras take part in hostile agendas, they do not do so by springing forth and getting in the way. They function through some other mechanism. They operate through some other hostile logic. My contention is that there are actually at least two "logics" by which cameras may participate in hostile agendas: (1) they extend the perceptual reach of hostile human actors and (2) they incite people under surveillance to police themselves.

The Logic of Confederacy

The first form of security camera hostility is perhaps the most straightforward: the camera transforms a potentially hostile viewer's perceptual capacity. That is, insofar as a camera (and associated surveillance system equipment) participates in a hostile agenda, one way in which it may do so is by extending the perceptual reach of hostile human actors. For example, imagine a hostile agenda which aims to deter a targeted population from using a particular public space and which employs human security guards to force them out. In such a case, security cameras could serve as tools to aid in the performance of this task. If the bench armrests and skatestoppers operate by a logic of physical imposition, then the cameras used in this manner operate by a *logic of "confederacy,"* acting as confederates to human actors enacting a hostile agenda. In this way, the cameras join other technologies that aid human actors in the enactment of hostile agendas, such as sign-in desks and other infrastructure that helps security personnel to manage public spaces.

This first form of security camera hostility—in which cameras work as confederates extending the perception of hostile human actors—can transform perception across different dimensions. Of course there is a transformation of spatial experience. The camera system transforms the spatiality of the user's perception. The camera system enables the security guard to monitor an area without herself or himself standing in that physical space. The camera may enable the area to be viewed from a perspective that could not be taken up otherwise. And of course a camera system can expand the range of what can be perceived. That is, a single guard may watch over the output from several cameras, monitoring several spaces at once. A camera may also enable a user to zoom in and magnify an area of interest in a way impossible for unaided perception.

But in addition to the multiple ways that a camera can transform the spatiality of a viewer's perception, cameras can also transform its temporality. As David Lyon puts it, "Whereas once surveillance in the city meant the use of street lights and physical architecture to contain deviance, it now also means keeping electronic tabs, including camera images, on the population at large" (Lyon 2001, 57). The person using the surveillance system not only may experience a transformed relationship to a particular landscape, she or he may experience a transformed relationship to the timescape of that location. Those under surveillance can be viewed not only in real time but a surveillance system may enable their history of action to be recorded, retained, and called up for viewing at a later time. Video recordings can be slowed down to watch the details of particular events or speeded up to scan for particular events taking place over a long period of time. And through the observation of patterns of behavior occurring within the space under surveillance captured on recording, it may be possible to better predict what behaviors will occur there in the future.

An important aspect of any instance of hostile design is how noticeable it is, what we could refer to as its level of "conspicuousness." Are instances of hostile design obvious to passersby? And to whom are they noticeable—to anyone, or only to those targeted? For example, nontargeted users of a public-space bench— that is, those who have only ever considered using it in its bench-as-seat stability— may not even notice the bench's anti-sleep features, like armrests or seat dividers. In contrast, anti-sleep features may be quite conspicuous to anyone looking to this public space for a place to rest for the night. Compare all of this to the example of the anti-homeless spikes. They are often quite noticeable, and noticeable precisely *as* hostile. In these terms, when a security camera is used as a perception-extending confederate, it often can best serve this function if the device itself remains inconspicuous. It may be easier to surveil a targeted population if they do not know they are being watched. The more inconspicuous the camera, the better it may serve its role as a hostile confederate.

But then why are public-space security cameras so often large, eye-catching, and surrounded by signage announcing that cameras are everywhere? This leads us to the second way in which security cameras can constitute a form of hostile design.

The Logic of Self-Coercion

The second form of security camera hostility is less straightforward than the first: the mere material presence of the camera itself may remind those in its vicinity that they should follow the rules, rules which may be written as part of a hostile agenda. If it is in the interest of a hostile agenda that the targeted be reminded that their behavior (or their very attendance) is unwanted, then the presence of security cameras may incite targeted people to change their behavior or to leave the area altogether. The visibly present camera thus can have the effect of enrolling the targeted into the very agenda that targets them.

In contrast to the anti-sleep benches or anti-homeless spikes that operate by a logic of physical imposition, or the first form of security camera hostility which operates by a logic of confederacy, this second form of security camera hostility operates by a *logic of self-coercion*. The visible security camera prompts the targeted to police themselves. In this way, the camera joins other self-coercive hostile designs, such as rule-listing signage, or even those plainly evident anti-homeless spikes, all of which remind the targeted population about the hostile agenda directed against them and encourage them to police their own behavior.

The work of Michel Foucault is of course a foundational contribution to the field of surveillance studies. Foucault combines these phenomenological insights into the experience of being surveilled with a radical account of the nature of power. Rather than understanding power only to be something that is possessed by one person and used on another—such as when a police officer cracks someone over the head with a baton—it can also be understood as something distributed across a system, shaping the context, infrastructure, and even the people within. Foucault's guiding example is the panopticon prison. In this style of prison, the prison cells are all part of a circular structure, facing inward, exposed to a guard tower in the center courtyard. At all times the prisoners can be viewed from the guard tower, and at all times the prisoners can see the tower from their cell. This constant surveillance, this constant knowledge that one's actions are subject to the perception of authorities, instigates one to follow the rules of that authority of one's own accord, to internalize that process, and thus to be shaped by it. Foucault writes, "He who is subjected to a field of visibility, and who knows it, assumes responsibility for the constraints upon himself ... he becomes the principle of his own subjection" (1977, 202–203). Foucault sees this mechanism broadly to be at work everywhere in our society, and more generally for selfhood to be shaped by the infrastructure of power. But for those studying contemporary surveillance, and especially those with a critical inclination, the application of Foucault's ideas is more direct. As Hille Koskela explains, just like the prisoner looking out at the guard tower, "citizens in urban space will see surveillance cameras positioned in visible places, and this will constantly remind them of their own visibility ... The panoptic condition of video-surveillance imposes self vigilance" (2000, 252–253). The visible security camera serves the function of the guard tower, and those under surveillance in public space are forced into the role of the prisoners.

This explains why public-space surveillance cameras are so often highly conspicuous. While a strategy of utilizing security cameras as part of a hostile logic of confederacy may favor inconspicuous or hidden devices, a hostile logic of self-coercion calls for making the surveillance apparatus itself as conspicuous as possible. It is of course common to not only see the cameras themselves conspicuously perched and dangling about but also signage advertising the fact that a space is under surveillance (Figure 7.5).

It is not necessary or even preferable that those under surveillance be made aware of the direction a camera is pointing. According to Foucault, it is not important that a guard be visibly looking in the direction of a prisoner, but only that the prisoner cannot see into the tower to know whether or not a guard is looking in his or her direction at that moment. The inability to see into the tower creates an ever-present threat of surveillance. Similarly, the hostile operation of security cameras in public space in terms of a logic of self-coercion may occur more smoothly and completely if the person under surveillance knows he or she is being watched but does not know exactly how and when. Signage pointing out that cameras are present can have this effect. Another common strategy is to

FIGURE 7.5 Parking lot signage in Boston, United States (photo by author).

obscure the current direction a camera is pointing by hiding it within an opaque dome or hanging globe.

And in Foucault's guard tower, it is not even the case that an actual guard must be present at all times, simply that the prisoners cannot see into the tower to know whether it is manned at the moment. Since for all they know a guard may very well be watching, the prisoners are impelled to self-police, thus maintaining "the automatic functioning of power" (Foucault 1977, 201). Similarly, not all security cameras need to even be turned on. And it is possible to purchase fake security cameras that simply provide the appearance of surveillance.

Even still, it remains possible for those under surveillance to grow accustomed to this condition, and to grow less and less explicitly aware of these dynamics. (In postphenomenological terms, we could say that one's relation to living under surveillance has grown transparent and sedimented.) The work of Benjamin Goold and colleagues points out this "everyday remoteness of camera surveillance, the fact that it is experienced as non-intrusive, discrete, part of the background, not as the foreground of social control" (2013, 948). At least for those not targeted by surveillance used as part of hostile agendas, the system of cameras and signage and other equipment—even when conspicuously present—can come to be ignored.

The Multistability of Surveilled Space

One important way that the example of the security camera is different from many of the other examples of hostile design is that its hostile operation changes a person's relationship to an entire space. Of course other hostile designs have this potential as well. But this is a key aspect of how we must conceive of the multistability at issue in security camera dynamics. In the case of a park or subway bench, it is the bench itself that is multistable, enabling seating and sleeping stabilities. The hostile design of an anti-sleep bench closes off one of these stabilities of the bench. The skatestoppers are similar. If they are installed on a handrail, then they function to close off one stability offered by the handrail itself, a skateboarding-related stability. The camera instead changes the targeted person's relationship to an entire space. In this case, the area under surveillance is the thing that is multistable. The camera (in the case of either hostile logic pertaining to the camera considered above) closes off a stability offered by that space. It contributes to an agenda of making the space available to some and not others. The effect is in some ways like that of those examples of hostile design that change the audioscape of public space in order to target a particular population. Those examples also change the composition of public space, discouraging particular uses of the space in a range of the audio devices.

Even still, the cameras themselves are also multistable in these examples in that they maintain different meanings within these different contexts. They can be interpreted differently depending on one's relation to them. For those targeted by a hostile agenda for which security cameras play a part, the cameras themselves may be experienced as significant, as difficult to ignore, as sources of authority whose gaze should be regarded wearily—and avoided if possible. We have noted that in contrast, the non-targeted populations of public spaces may instead experience security cameras as simply part of the forgettable backdrop.

This points to a direction for resistance efforts. Many of these efforts, especially those of activist artists, put a spotlight on the pervasive surveillance apparatus all around us. This kind of work is important, especially for highlighting concerns that might be shared by a majority of people, such as the defense of privacy. But the conception of public-space security cameras as potential instances of hostile design suggests another avenue for resistance: the exposure of cameras specifically in terms of their roles in hostile agendas. We should hold up and amplify resistance works that not only expose the infrastructure of surveillance but call out and criticize it *as* hostile, that is, *as* contributing to concerted efforts to target and discriminate against specific groups. Support for targeted groups must be marshalled, and alternatives to hostile treatment built into our infrastructure must be developed.

In this way, consciousness-raising efforts regarding issues of surveillance should not reduce to those concerned with protecting privacy. Those who have been historically targeted by discriminatory surveillance campaigns, such as immigrant communities, people living in poor urban neighborhoods, Muslims in the West, and LGBT folks, have not always been afforded privacy to begin with. This is especially salient for those living unhoused. For those without private space, for those living their lives in public, privacy is already denied. A key thing to highlight about surveillance, then, is not simply how it robs the unhoused of a right to privacy. We must expose the roles of surveillance devices in the larger agendas to push the unhoused and others out of public spaces, out of safety, out of the community, and out of view.

Perhaps the concepts considered above can be useful in drawing out the ways that technological mediation is often a relation of control. And perhaps they can help us to recognize our complicity in unjust networks of control, to develop forms of resistance, to build new networks of solidarity, and to imagine new possibilities for the things of our world.

Notes

1   While here I only draw on Akrich's work for the sake of streamlining this account of the potential combination of postphenomenology and actor-network theory, elsewhere I have followed this out in more detail, especially in consideration of the intricacies of Bruno Latour's actor-network account of technology (e.g., Latour 1999). See, for example, Rosenberger (2014, 2017a, 2017b, 2018).
2   While I do hold that postphenomenology and actor-network theory can provide the basis of a useful account of hostile design, I also recognize that these theories do not by themselves provide much to capture the politics of this topic with any precision. Elsewhere I attempt to connect this account up with political theories of technology, especially critical theory and feminist standpoint epistemology (Rosenberger 2017a).

References

Akrich, M. 1992. "The De-Scription of Technical Artifacts." In *Shaping Technology/Building Society*, edited by W.E. Bijker and J. Law, 205–224. Cambridge: MIT Press.

Ball, K., K. Haggerty, and D. Lyon, eds. 2012. *Handbook of Surveillance Studies*. New York: Routledge.

Chellew, C. 2016. "Design Paranoia." *Ontario Planning Journal* 31 (5): 18–20.

Davis, M. 1990. *City of Quartz: Excavating the Future in Los Angeles*. London: Verso.

Doherty, J., V. Busch-Geertsema, V. Karpuskiene, J. Korhonen, E. O'Sullivan, I. Sahlin, A. Tosi, A. Petrillo, and J. Wygnánska. 2008. "Homelessness and Exclusion: Regulating Public Space in European Cities." *Surveillance & Society* 5 (3): 290–314.

Flusty, S. 1994. *Building Paranoia: The Proliferation of Interdictory Space and the Erosion of Spatial Justice*. West Hollywood: Los Angeles Forum for Architecture and Urban Design, Forum Publication No. 11.

Forss, A. 2012. "Cells and the (Imaginary) Patient: The Multistable Practitioner-Technology-Cell Interface in the Cytology Laboratory." *Medicine, Health Care & Philosophy* 15: 295–308.

Foucault, M. 1977. *Discipline and Punish*. New York: Vintage.

Friesen, N., A. Feenberg, and G. Smith. 2009. "Phenomenology and Surveillance Studies: Returning to the Things Themselves." *The Information Society* 25: 84–90.

Goold, B. 2004. *CCTV and Policing*. Oxford: Oxford University Press.

Goold, B., I. Loader, and A. Thumala. 2013. "The Banality of Security: The Curious Case of Surveillance Cameras." *British Journal of Criminology* 53: 977–996.

Hasse, C. 2015. "Multistable Roboethics." In *Technoscience and Postphenomenology: The Manhattan Papers*, edited by J.K.B.O. Friis and R.P. Crease, 169–188. Lanham: Lexington Books.

Ihde, D. 2009. *Postphenomenology and Technoscience: The Peking University Lectures*. Albany: SUNY Press.

Koskela, H. 2000. "'The Gaze without Eyes': Video-Surveillance and the Changing Nature of Urban Space." *Progress in Human Geography* 24 (2): 243–265.

Latour, B. 1999. *Pandora's Hope*. Cambridge: Harvard.

Léopold, L. 2013. *Cruel Designs, The Funambulist Pamphlets, Vol. 7*. New York: Punctum Books.

Levin, T.Y., U. Frohne, and P. Weibel, eds. 2002. *CTRL Space: Rhetorics of Surveillance from Bentham to Big Brother*. Cambridge: MIT Press.

Low, S. 2003. *Behind the Gates: Life, Security, and the Pursuit of Happiness in Fortress America*. New York: Routledge.

Lyon, D. 1994. *The Electronic Eye: The Rise of the Surveillance Society*. Cambridge: Polity Press.

Lyon, D. 2001. *Surveillance Society: Monitoring Everyday Life*. Buckingham: Open University Press.

Marx, G.T. 2015. *Windows in the Soul: Surveillance and Society in the Age of High Technology*. Chicago: University of Chicago Press.

Minton, A. 2012. *Ground Control: Fear and Happiness in the Twenty-First-Century City*. London: Penguin.

Mitchell, D. 2014. *The Right to the City: Social Justice and the Fight for Public Space*. New York: Guilford.

Monahan, T. 2010. *Surveillance in the Time of Insecurity*. Piscataway: Rutgers University Press.

Newman, O. 1973. *Defensible Space: Crime Prevention Through Urban Design*. New York: Collier Books.

Norris, C. and G. Armstrong. 1999. *The Maximum Surveillance Society: The Rise of CCTV*. Oxford: Berg.

Petty, J. 2016. "The London Spikes Controversy: Homelessness, Urban Securitisation and the Question of 'Hostile Architecture.'" *International Journal for Crime, Justice and Social Democracy* 5 (1): 67–81.

Quinn, B. 2014. "Anti-Homeless Spikes Are Part of a Wider Phenomenon of 'Hostile Architecture.'" *The Guardian* 6 (13). https://www.theguardian.com/artanddesign/2014/jun/13/anti-homeless-spikes-hostile-architecture.

Rosenberger, R. 2009. "The Sudden Experience of the Computer." *AI & Society* 24: 173–180.

Rosenberger, R. 2014. "Multistability and the Agency of Mundane Artifacts: From Speed Bumps to Subway Benches." *Human Studies* 37: 369–392.

Rosenberger, R. 2017a. *Callous Objects: Designs Against the Homeless*. Minneapolis, MN: University of Minnesota Press.

Rosenberger, R. 2017b. "On the Hermeneutics of Everyday Things: Or, the Philosophy of Fire Hydrants." *AI & Society* 32: 233–241.

Rosenberger, R. 2018. "Why It Takes Both Postphenomenology and STS to Account for Technological Mediation: The Case of LOVE Park." In *Postphenomenological Methodologies*, edited by J. Aagaard, J.K.B. Friis, J. Sorenson, O. Tafdrup, and C. Hasse, 171–198. Lanham: Lexington Books.

Rosenberger, R. (2019). "On Hostile Design: Theoretical and Empirical Prospects." *Urban Studies*. https://doi.org/10.1177/0042098019853778

Rosenberger, R. and P.-P. Verbeek, eds. 2015. *Postphenomenological Investigations: Essays on Human-Technology Relations*. Lexington Books/Rowman Littlefield Press.

Rosenfeld, A. 2015. "Mediating Multiplicity: Brain Dead Bodies and Organ Transplant Protocols." In *Postphenomenological Investigations*, edited by R. Rosenberger and P.-P. Verbeek, 203–214. Lanham: Lexington Books.

Savivic, G. and S. Savic, eds. 2013. *Unpleasant Design*. Belgrade: G.L.O.R.I.A.

Schindler, S. 2015. "Architectural Exclusion: Discrimination and Segregation Through Physical Design of the Built Environment." *The Yale Law Journal* 124: 1937–2024.

Van Den Eede, Y., S.O. Irwin, and G. Wellner. 2017. *Postphenomenology and Media: Essays on Human-Media-World Relations*. Lanham: Lexington Books.

Verbeek, P.-P. 2011. *Moralizing Technology: Understanding and Designing the Morality of Things*. Chicago: Chicago University Press.

von Silva-Tarouca Larsen, B. 2011. *Setting the Watch: Privacy and the Ethics of CCTV Surveillance*. Oxford: Hart Publishing.

Wellner, G.P. 2016. *A Postphenomenological Inquiry of Cell Phones: Genealogies, Meanings, and Becoming*. Lanham: Lexington Books.

8 A TOOL FOR THE IMPACT AND ETHICS OF TECHNOLOGY: THE CASE OF INTERACTIVE SCREENS IN PUBLIC SPACES

Steven Dorrestijn

Introduction

In this chapter I will elaborate on the relationship between the impact and the ethics of technology, and more specifically on the significance of the so-called Product Impact Tool for ethical reflection and discussion about technology. A case study concerning ethical reflection in a research project about the development of interactive screens in public spaces is integrated as an illustration and application of the approach.

The Product Impact Tool issued from research aiming to make the connection between philosophical reflection on technology and the design of technology (Dorrestijn 2012, 2017a). The core of the tool is a framework that offers a practical translation of the notion of technical mediation (that human existence and way of living are fundamentally mediated by technology) by breaking the general idea down into four modes of interaction, and twelve types of impact, illustrated with examples. It is thus a collection of exemplary figures of technical mediation against the background of the general idea.

When applied in design for usability or behavior change, the tool is related to "affordances" in design (Norman 1988), "persuasive technology" (Fogg 2003), and "nudge" (Thaler and Sunstein 2008), three approaches referred to in the tool. However, an important application of the Product Impact Tool is to support and stimulate ethical reflection about technology and behavior-changing design. From the beginning responsible innovation has been a goal of the tool. The tool, however, does not explicitly employ ethical terminology but only speaks of impact. This raises the following questions: How does the Product Impact Tool, in the practice of an impact assessment with the tool, encourage ethical reflection? And how can the conceptual connections between the impact and ethics of technology be understood, in order to explain the passage from impact to ethics, and to structure and enhance the use of the tool for ethical reflection?

The chapter proceeds as follows. The Product Impact Tool is shortly introduced in relation to the philosophy of technical mediation. Then I report on the project about interactive screens in public spaces where the tool was applied in a workshop leading to a concise advice on ethical aspects. Next, I elaborate how a technology impact assessment with the tool and ethical reflection about technology are correlated. To begin, it appears that in practice impact assessment workshops appear to result in ethical reflection and discussion. Then I proceed by further developing the experience that impact assessment leads to ethical reflection also in conceptual way. By means of examples of impact of interactive screens brought to the fore in the workshop, the different modes of interaction and impact are introduced. From there I explore how every figure of technical mediation in the tool is at the same time a concept for understanding impact as well as an expression of ethical concern. The impact of technology often alludes to a negative value and engenders critical concern. Any type of impact can, however, also take on positive value and then a design strategy and ethical remedy can be derived.

In terms of this book, this chapter is about how we can or should relate to things (ethical concern about technology), especially also in response to how those things relate to us (impact of technology). With the case of interactive screens and the focus on impacts and ethics, the issue of control comes to the fore as an important aspect of the relation between things and us. Technologies have tendency to control us, and interactive technologies may come to control us in new and more intensive ways. It is a challenge of our time to become aware of the controlling tendencies of technology. And a key ethical question concerns the meaning of this control by technology as well as which kind of control over technology we wish to have.

Product Impact Tool: Figures of Technical Mediation

The Product Impact Tool (Figure 8.1) provides an overview of different effects of technical products on human behavior and existence. It is an interdisciplinary collection of relevant examples and useful concepts from a variety of scholarly disciplines, from design to psychology, history, anthropology, and philosophy. The central model shows a human figure in the middle of a repertoire of types impact, thus expressing how technology is taking hold of humans from all sides. The twelve types of impact are subdivided into four different modes of contact and interaction. This repartition aims to cover the impact of technology in full scope while remaining sufficiently comprehensible for memorization and application.

In the online version of the tool (see www.productimpacttool.org), one can unfold the framework from the diagram in the center, via text boxes with

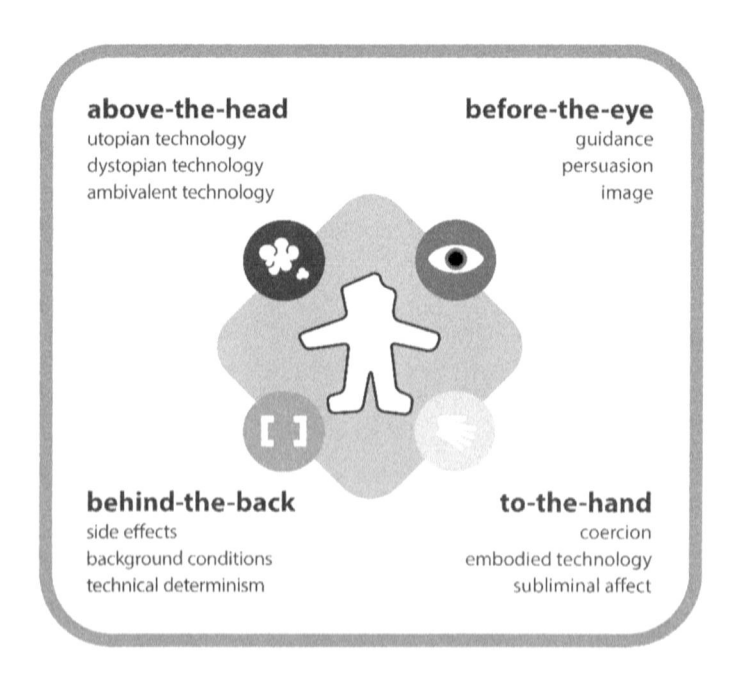

above-the-head
utopian technology
dystopian technology
ambivalent technology

before-the-eye
guidance
persuasion
image

behind-the-back
side effects
background conditions
technical determinism

to-the-hand
coercion
embodied technology
subliminal affect

FIGURE 8.1 Product Impact Tool Model.

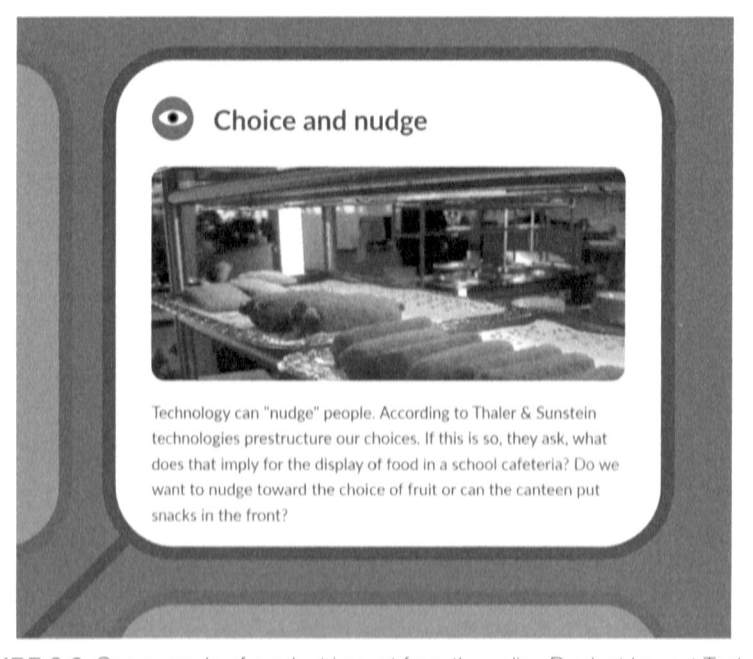

Choice and nudge

Technology can "nudge" people. According to Thaler & Sunstein technologies prestructure our choices. If this is so, they ask, what does that imply for the display of food in a school cafeteria? Do we want to nudge toward the choice of fruit or can the canteen put snacks in the front?

FIGURE 8.2 One example of product impact from the online Product Impact Tool.

explication (Figure 8.2) of the modes of interaction and types of impact, to the outer circle consisting of sixty examples with a picture and a short explanation.

The Product Impact Tool intends to give a practical turn to work in the philosophy of technology (Eggink and Dorrestijn 2018). It refers especially to the current of historical, empirical, and phenomenological research into "technical mediation" (cf. McLuhan 2003 [1964]; Ihde 1990; Latour 1994; Verbeek 2005). The term "mediation" is conveniently broad so that it can serve, in my view, as an overarching concept for how technologies play a role in, intervene in, and give a twist to our existence and way of living. It is an important element of the concept of technical mediation that technology and humans are not approached as fundamentally separated entities, but rather there is a focus on interactions, interdependencies, or interwovenness. The focus in the tool is on the impact of technology, but the *impact of technology on humans* is framed within the *interaction between humans and technology,* and overall, the model reflects the idea of *technical mediation of human existence.*

Within the framework of technical mediation, the Product Impact Tool brings together a variety of answers to the question what technologies do to our existence, as have been discovered, formulated, and conceptualized by people in different times and from different disciplinary angles. It is not the ambition to provide one single answer to the question what technical mediation is and how it works. The relations between humans and technology appear so complex and dynamic that striving for a complete grasp (in the form of some quasi-mechanistic theory of human-technology relations) seems unfeasible. Rather my approach is to explore the multiplicity of effects of technology that may occur and provide some structure by the framework of modes of interaction and types of impact. The Product Impact Tool model thus collects conceptions of specific "figures of technical mediation," appearing against a "background" of technical mediation as a general theme or problematic (cf. Dorrestijn 2012, 62–64).

The resulting repertoire of figures of technical mediation can be applied for several purposes. The tool can be of help in the design process, supporting design for usability and for behavior change. Also the tool can help to analyze problems with the acceptance of technologies. And, as the focus is the present text, the tool also contributes to the ethics of technology.

Impact and Ethics of Interactive Screens in Public Spaces

A concrete example which I will use as a case in this chapter is offered by the research project OBSERVE, in which interactive screens for public spaces are developed (Figures 8.3 and 8.4). Big screens are popping up everywhere: alongside highways, on squares, on the facades of shops and schools. OBSERVE is about making the content on those screens interactive: responsive to circumstances such

FIGURE 8.3 Project name display on a screen on a pilot day. Source: http://www.actmedialab.nl/tweede-pilot-observe/.

FIGURE 8.4 People in view of an interactive screen system. Source: http://www.actmedialab.nl/tweede-pilot-observe/.

as the weather, or festivities, or also responsive to input from people passing by or to information gathered about them by sensors and from the internet. A business incentive for this project is that responsive content would render the exploitation of screens more economical because of better targeting of people. A simultaneous promise is that content will become more relevant and enjoyable for the public.

To make the interactivity work, such a system makes use of sensors, as well as the collection of information from data banks and from the internet. The collected information is used real time by the system for selecting relevant content or to produce content through input from the spectators.

A consideration of the ethical aspects of interactive screens was part of the research project from the offset. Initially the ethical aspects were framed in a rather juridical way: What does data and privacy legislation allow concerning the collecting and use of data about people, and does the project remain within these borders of law? From the perspective of the Ethics and Technology research group it seemed immediately that the ethical issues were broader and different from the question about compliance with privacy legislation. The pilots in the research project would not immediately be so advanced that privacy laws would be an important factor—not yet. Still, the future of interactive screens, especially the possibility of personalization of content by responsiveness of the screens to specific people in front of the screen, may have massive consequences with respect to privacy. Many people are reminded of the scene in the movie *Minority Report* (2002, directed by Steven Spielberg), where one of the main characters (played by Tom Cruise) walks in a site where all the walls are screens which show messages specifically directed at him. In such a fully interactive environment the data collected about people returns to them in the form of interference in their behaviors in very personal and far-reaching ways.

Interactive environments may prove altogether impossible to exist if current privacy principles of informed consent about data collection, storage, and processing would be strictly followed. Moreover, people themselves share data on social media and other online services, while consent is only a formality, but in no way can be actually considered informed consent. And how could this be different? If full data transparency will be impossible in a future smart world, what does this mean? Total data transparency is hardly feasible, but also not what most people seem to want. It is improbable that technical developments will be put on hold, nor that many people would really want full abandonment of interactive technology. Are we therefore seeing the end of privacy? There is another option. It may also be that the meaning given to the value of privacy is transforming. Even if all data streams can no longer be controlled, people will remain or become more and more cautious about the meddling of smart technologies in their lives, the practical interference with their freedom. Following this approach, emphasis in the framing of the problem is thus shifting from data to impact. In dealing with impact, the framing is also shifting from a focus on regulation by law to a focus on responsible research and innovation.

Fitting with this orientation toward responsible research and innovation, an important activity in the project OBSERVE was a workshop with project group members about the impact and ethics of interactive screens (on July 12, 2016, with six research group members, and myself as workshop leader). The approach followed was to start by analyzing the impacts of interactive screens in a structured way with the help of the Product Impact Tool. The model was presented using the online version of the Product Impact Tool. Each time the three types of impact in one of the quadrants were presented, and then the group discussed if these exemplary effects could be recognized or imagined to apply in the case of interactive screens in public spaces.

The details of the discussions in the workshop will be reported on hereafter in the elaboration of relationships between figures of impact and ethical concern. I will end this section with a concise reproduction of the five points of attention which were drafted by way of conclusion in the workshop proceedings.

Awareness and Responsibility

The workshop was a lively event. The introduction to the workshop suggested issues for reflection, but participants also brought their own knowledge, intuitions, opinions, and topics to the discussion. The workshop offered occasion for articulation and deliberation of one's thoughts. The workshop did help to raise awareness, and this confirms the assumption that not only legal compliance but also responsibility of participants is important in such a project.

Respect Existing Legal Regulation

Compliance with data and privacy legislation is of course important and juridical advisors are needed for this. But there are many gray areas and novel situations where existing law is not adequate so that ethical reflection and responsibility are impelled.

Privacy and Interference

Framing the ethical aspects broadly as concerning freedom and interference with behaviors, beyond data and privacy, appeared mostly a fruitful approach. The theoretical introduction about this point was taken for granted without any special approval or disapproval, but the topics discussed during the workshop indeed covered this broader thematic of impact and freedom.

Participatory Design

At several occasions during the workshop the project became characterized as *top down*. The desire of the public for interactive and personalized content would be largely an assumption from developers. People in the street have a very different estimation of technical opportunities; most of them are not early adopters. The acceptance readiness of people does not, for the most part, follow from arguments or technical numbers but is based on people's feelings. This calls for communication between developers and users, like in a product impact workshops where opinions and emotions can become articulated and discussed.

Participatory Interaction

Participatory design means in a minimal sense that users are heard and informed in the design process but can in a broader sense also mean that the functionality actually is affected by the input from users. The understanding of interactive technology by developers often seems to have the form of a sensor feedback control system, where data about users is gathered (by sensors or by data mining in available databases) and users remain passive. The possibility of active contribution by users (by input via speech, gesturing, pushing a button) remains underexposed. Still this may offer new possibilities in terms of interaction, which are highly interesting because of a real interplay between humans and the technical system. At the same time in the case of such participatory interaction ethical issues with privacy and behavior interference will often be circumvented or prevented. For when people retain an active role *in the loop*, they stay more in control.

Correlating Figures of Impact and Ethics

The aim of this text is to elaborate how the Product Impact Tool can also serve as a tool for ethical reflection and discussion about technology. To begin, a structured impact assessment raises awareness of otherwise unnoticed or neglected effects of technology. In a Product Impact Tool workshop, like the workshop on interactive screens, this usually leads to much discussion and this brings along the articulation of ethical concerns about technology. This can clearly be recognized in what is reported in this text on the workshop about interactive screens. Therefore it appears that the Product Impact Tool already functions as a tool for stimulating ethical questioning, even when the terminology used in the tool is about impact and does not explicitly refer to ethics.

It is still, however, an interesting question how this pattern of ethical discussion raised by assessing impacts can be explained and structured. In this section I will therefore review the correlation between figures of impact and of ethical care, as two sides of the same coin. While the recognition of impact may produce ethical concern, the conceptualization of impact can just as well be seen as an expression of ethical concern. It may well be that often the concern came first and the conceptualization of effects of technology followed in reaction to the concern. This reflects the epistemological issue that knowledge about ourselves, and thus also about technology affecting us, cannot be altogether objective (cf. Dorrestijn 2017b).

In the following, the different types of impact and the associated ethical concerns will be discussed, starting from the discussions in the interactive screens workshop. This means a further presentation and explanation of the types of impact in the Product Impact Tool with illustrations from the case of interactive screens. The same workshop discussions are also the starting point for the elaboration of the correlated figures of ethical care.

The mirroring between the perspectives of analysis of impact and of ethics is most obvious in the above-the-head quadrant of the tool, which comprises overarching philosophical and ethical visions on the impact of technology (utopian, dystopian, and ambivalent technology). But in some of the other exemplary mediation effects, the ethical aspect is also readily recognizable. Side effects and technical determinism are good examples. These notions have a critical connotation, alluding mostly to undesired impact—although they surely also can take on a positive value. Embodiment and guidance, by contrast, are types of impact which may have predominantly a positive ethical connotation. I will start behind-the-back and conclude above-the-head.

Behind-the-Back

Behind the visible and tangible screens themselves, interactive screens assume much technical and organizational infrastructure in the background. Think of sensors, archives with content to be broadcast, a content selection system, an exploitation and business model, etc. The impact of technology behind-the-back affects people indirectly, without direct user-technology contact. The *behind-the-back* mode of interaction appears particularly important in this case, as with regard to intelligent environments in general.

Background Conditions

The functioning of interactive screens is in numerous ways dependent on the wider infrastructure. Considering *background conditions* often helps explaining problems with the implementation and adoption of new technology. During pilots in this project about interactive screens, this appeared an important dimension. Fast internet connection proved to be a bottleneck for the fluent functioning of the system. The geographical site also determines essential environmental factors of the system. Think of the amount and variety of people passing; weather circumstances; and the presence of historical, touristic, or commercial spots. The content selection mechanism is a clear example of a system that functions in the background, which users cannot directly interact with or even see, but which is very important with respect to ethics.

Background conditions tend to be concealed and often mean dependence, making it one of the impact figures with a mostly negative ethical connotation. For, uncontrollability contradicts the ethical ideals of awareness and consent. Still, withdrawal to the background can also assume a positive sense as it also means absence of hindrances, convenience. In this sense background conditions, persuasion, subliminal effect, and embodiment are all comparable. The ethical remedies to the concerns about background conditions are *raising awareness* and a systemic design approach for improving *integration* of different factors. Redesign of a whole system becomes in the strongest form a *regime shift* or *Revolution*, as in Marxism, where political and cultural transformation would be achieved by changing the material and economic basis of society.

Side Effects

Next to considering interdependences of background conditions, a second form of impact behind-the-back concerns *side effects*. One example from a pilot in OBSERVE during spring time was that movie clips appeared on the screen about Saint Nicolas festivities, which one would only expect in the weeks up to December 5. The automatic selection of content led to unintended consequences. Another side effect is that while the screens aim to draw people's attention to the screens, they will distract them from everything else, which can pose serious safety risks in traffic for example. Other unintended consequences are the visual pollution during daytime and undesirable spoiling of darkness at night of all-too-bright screens (so considered by many, although a matter of taste).

Clearly the impact figure of *side effects* is also one with negative ethical qualification. This is true for all the examples discussed in the interactive public screens workshop: visual pollution, distraction of people, and automatically selected content that is not fitting. Other terms related to side effects, such as *unintended consequences*, *collateral damage*, or *risks*, all sound negative. Even when often negative, side effects can also occur in a positive way; think of the notion of a *win-win situation*. In the normal sense where side effects are mostly unintended and unforeseen, the corresponding ethical remedy is anticipation and prevention of the negative impact.

Technical Determinism

The third behind-the-back type of impact that the Product Impact Tool distinguishes is *technical determinism*. This term refers to a philosophical and ethical question: Do humans design and control technology, or is it rather an internal logic and power of technology which determines human culture? But it can also be considered in a descriptive way and on a smaller and more concrete scale: Does technology offer solutions to existing needs and values, or does the availability of technology create or change needs and values? In the case of interactive screens an important aspect is what happens to the value of privacy. Awareness and consent about the use of personal data seem almost impossible with rush of smart and interactive technology. Must privacy be reinforced, redefined, or can only its demise be lamented? Another instance of *technical determinism* which is relevant here is what is called *function creep*. Can it be prevented that personal data, first collected solely for monitoring movements in the city, will later perhaps be used for personalized publicity or for regulating access?

In the explanation of *technical determinism* as technology producing instead of following human needs and values, the ethical significance is already quite explicit. Function creep denotes an ethically undesired form of changing needs. The challenging of values, such as privacy, has also a predominantly negative meaning. However, in the ethics of technology it is an important question if there is a middle way between acknowledgment of the impact of technology on human values and leaping into full determinism (Swierstra, Stemerding, and Boenink 2009; Kudina and Verbeek 2019). The ethical concern and remedy with respect to value change are aptly denoted by the quite popular notion of *disruptive technology*. The term is often used by technology enthusiasts in a positive way expressing that new technology enforces a break out of a dated regime (like Uber disrupts an overregulated and protected taxi system). Still, in ordinary language disruption rather used to have a negative ethical connotation. Much depends on the values one endorses. In negative disruption the idea of a decline of human values, overruled by technical change, takes prominence. Positive disruption converges with regime change, as in changing the background conditions for a good cause. The change of values may in this case be seen as an awakening (an overcoming of false ideology). The core ethical significance about such a challenging of values is not the empirical description of changing values, nor philosophical acknowledgment that values may change, but critical reflection about the question which values, long standing or revaluated, we want to affirm.

Before-the-Eye

The quadrant *before-the-eye* is about technology as carrier of information and meaning, addressing the user's cognition and influencing people by informing their decisions for actions. In the case of interactive screens this category comprises the content shown on screens but also the appearance (design) of the screens.

Guidance

A first point which was discussed is whether people were at all able to see and understand how the interactive screen reacts on their presence, and how they can influence the interaction. This is an example of the effect of *guidance* in the Product Impact Tool: design elements which provide guidance or information about the intended use of technology.

Guidance has a mostly positive value, which is easily translated into a design strategy. For example, the usability expert Donald Norman (1988) with his work on affordances has promoted user-guiding design as an important strategy for user-friendly design. His approach is a response to annoyance with the opposite, of technologies which are *misleading* users. User guiding design can therefore be a design strategy and serve as an ethical remedy. Another (positive) ethical significance of guidance is that it complies with the ethical value of being informed.

Image

The design of a screen must also fit the city and the specific place. A workshop participant brought up the example of a new digital time table at the Central Station of the Dutch city of Utrecht. Many people, however, long for the former analog time table (with flipping slats). The style and branding of a technology matter for the acceptance. Do we or do we not want to associate or identify ourselves with certain technologies of a certain style or brand, or a certain kind of technology at all? In the Product Impact Tool this falls under the category of *image*. Regarding the question of image, the research project title, OBSERVE, is quite daring. The reference to observation will remind people of issues with privacy and control (Big Brother), and this is perhaps not a lucky association.

The notion of *image* can just as well be positive as negative (like the title OBSERVE). A negative image that proves persistent becomes a *stigma*. A different kind of ethical concern about image is that it is *only superficial*.

Persuasion

The effect of *persuasion* (cf. Fogg 2003) is a more intentional kind of influencing by technology than just conveying information or meaning (*guidance*). It is definitely relevant with regard to the content displayed on the screens. It is a prerequisite that the screen itself is *persuasive*, in the sense that it must first successfully attract the gaze of people. The consideration of persuasion led one of the workshop participants to remark that up to this point content makers for public screens are very much focused on broadcasting with the goal of persuading and influencing people, and not on true interaction (including a participatory role for the public).

The ethical concerns about *persuasive technology* are especially complex and interesting. First, persuasive technology and the comparable concept of *nudging* (Thaler and Sunstein 2008) are typically used for ethically desirable causes, such as saving energy and stimulating healthy eating and lifestyle. The idea is that persuasive technology functions as behavior support to tune behavior toward goals we ourselves affirm but often fail to achieve without the right cues. Second, the theories of persuasive technology and of nudging both include ethical guidelines where the idea of informed consent is important and where the targeted people should always be able to opt out. However, these ideals can easily become challenged in practice. For who decides what are good causes? And cannot the same strategy be used by people with bad intentions?

Moreover, while the theory says explicitly that there should be awareness about persuasive or nudging strategies, the effectiveness might be better when there is less awareness on the user side. Think of hidden messages, a related technique which influences people's decisions subconsciously. It is impossible to prevent that this technique is used for online marketing or influencing elections via social media, but it is fully against the transparency guideline.

Finally, even when the persuasion is transparent and fully in accord with the ethical guidelines, there remains a fundamental ethical issue. Take as an example a persuasive message on a screen in a public space: receive a coffee for free when you park your car outside the city center, and keep the city clean and pedestrian-friendly. It is an ethical problem (especially in Kantian ethics of good intentions, duty, and autonomy) that people then do good, but only because of something else (a coffee for free) and not because of intrinsic moral motivation.

The downsides of pervasive augmented reality and persuasive technology are impressively explored in the artistic movie *Hyper Reality* (by Keiichi Matsuda; see http://hyper-reality.co/), which is discussed extensively by Galit Wellner (see Chapter 9). The sheer amount of images and messages in this example of *augmented reality* is completely overwhelming for the main character of the movie and for us as spectators. When, on top of this, the system is hacked, the situation gets even worse and really confusing.

To-the-Hand

In the Product Impact Tool *to-the-hand* denotes physical interaction by which products affect people directly by interference with their body and gestures. In the case of interactive screens this is not the most important quadrant, but still there are effects which have to do with the materiality and the positioning of the screens and with the physical sensations of the light of the screens.

Coercion

The screens must be placed so that people do find them. Are they well visible, or too obtrusive? Even if it is not the typical example of *coercion*, it can be said that a screen which cannot be practically circumvented does physically *coerce* people's gaze toward the screen and confront them with the content.

Coercion has an obvious ethical counterpart in the *obtrusion of freedom*. And this means, following the famous analysis by Latour (1994), that morality is taken from humans and delegated to things. Coercive design can bring along usability annoyances but can also go as far as a total control of behavior, which is obviously ethically significant. Interventions such as spikes to ward off the homeless or youths with skateboards are a good example (see Rosenberger in this volume). This also gives insight in yet another aspect, namely that because the constraints are often both unwanted and well visible to the targeted people, they may provoke subversive action. Finally, a positive ethical value of coercive design is that there is a fixed procedure, and users cannot do anything wrong.

Subliminal Affect

Not only the positioning but also the often-extreme brightness of some screens is obtrusive if one passes by in the darkness of the night. The light can be so bright that it hurts your eyes, resulting in a repulsive reaction. This is an example of *subliminal affect*, where the senses (sight in this case) are addressed not with information to be cognitively processed but in a sheer physical way, rendering an effect of attraction or repulsion.

Subliminal affect was found to be somewhat applicable in the case of all-too-bright screens, but it should be noted that such light does reach the level of conscious cognition. The subliminal aspect comes more to the fore when for marketing purposes, for example, people's moods are being influenced with light, color, sound, or smell. This prompts an important ethical challenge, because people remain mostly or altogether unaware about how they are influenced: the opposite of the ethical ideal of informed consent.

Embodiment

Public screens are a simple technology in the sense that little learning efforts are needed to be able to watch them. They are easily *embodied* (although that is on the basis of knowledge and skills to understand the place of screens and images and text in our world, which everybody must acquire through education). But embodiment may acquire extra relevance when the interactivity of such screens is further developed. In a project by artists in the research group, the movements of people on the square in front of the screens were monitored and used to generate a line on the screen. People are thus invited to enter in an interplay with the screen and start to adapt their movements on the screen in order to influence the drawing on the screen. This is an example of the possibility of human-technology interaction by movements and gestures. This is a trend: think of swiping on touch screens or contactless controlling by gesturing of gaming consoles. It means a retrieval of the relevance of the impact figure of *embodiment*, albeit in a new form. This time it is not about literally handling tools but about remote-controlling technical systems by gesturing.

In general, *embodied technology* has a positive ethical connotation. In interaction design it is a mark of successful design when a product is easy to use, without thought, in a natural way. It is undesirable when the learning curve is too steep, because a technical product demands much "use technique" (skills) (Tenner 2003). The aspect of habituation and training of embodiment implies a self-transformation of the user. For this reason embodiment is an important notion in the ethics of technology when the focus is on material culture, ethical practice, and self-development. This practical focus stands in contrast with a more theoretical and cognitive understanding we have about our behavior and ethics where the focus is on conscious behavior and technology use. An ethical concern about embodied technology, as with the other subconscious figures of impact, is that it goes against the ethical ideal of consent. This concern becomes a concrete danger when the attachment and interaction become so easy, natural, or pleasant and people may slip into a form of addiction to technologies. The initial burden of habituation reverses to a burden of rehab or detox. This is at least the terminology that is being used more and more in relation to the pervasive use of the smartphone and social media apps.

Above-the-Head

Finally, the *above-the-head* quadrant comprises overarching ideas about the impact of technology. In the case of interactive screens, the views concern typically expectations about perfect automation and personalization as well as concerns about the values of privacy, freedom, and control in public space. Whereas impacts in the other quadrants refer to concrete examples, here we find generalizing, abstract views on the meaning of technology for humans. The tool contains three conceptions of technology (utopian, dystopian, and ambivalent technology) as a very concise and schematized overview of the philosophy of technology. Even if there is no explicit reference to ethics in these titles of the figures of mediation, they do express an ethical valuation (from altogether positive, to ultimately negative, to mixed). Moreover, the more extended explanations of the views on technology in the online version of the model and in background literature (Dorrestijn 2012, ch. 4; 2016, 2017b) do explicitly combine an analysis of the impact and an evaluative, ethical counterpart.

Besides the ethical concern as counterpart of the impact, it is also possible to make a connection between understanding of impact and ethical theories and principles. The utopian view on technology lines up mainly with utilitarian ethics (after Jeremy Bentham), while the dystopian view rather combines with duty ethics (Immanuel Kant). Both of these moral theories are rationalistic and theoretical. The view of ambivalent technology is more congruent with practical currents in ethics, with a focus on virtues and on care of the self (Michel Foucault). I have elaborated these links between conceptions of technology and moral theories elsewhere (see Dorrestijn 2012, ch. 5) and will include very concise summaries in the following.

Utopian Technology

In the workshop it was remarked that the screens are being imposed upon the public. The screens have come into being more due to technology push than marked pull. This links to the figure of *technical determinism* in the sense of the history and governance of technology. But it was also discussed how this pushing of technology has to do with the meaning given to technology, a positive attitude on the side of developers. The enthusiasm can have traits of a *utopian view of technology* and is often focused too much on technical possibilities and solutions while neglecting actual human preferences and values.

My description of utopian technology, including the ethical concern, is *that technology wonderfully completes human life, while the ethical challenge is only to solve scarcity or unequal distribution of technology*. Technology itself here appears good; only the application and distribution, an economic problem of a good fit to human needs, can be problematic.

The economic perspective is also central in the ethical theory of utilitarianism, of which Jeremy Bentham (1748–1832) is a main proponent. Bentham stressed that utility should be the principle of a radically rational ethics. Good is an action which results in maximum happiness (for oneself and others, for the greatest number). Interestingly, Bentham did explicitly consider technology, as he was also an avid promotor of his Panopticon project, an architectural design for a prison or any building for holding together large numbers of people. The circular shape with a central watchtower allows for continuous inspection. This effectively prevents any incorrect behavior but also, as Bentham believes, removes the will to do evil. In the actual world actions that go against the principle of promoting happiness for the community may go unpunished or even prove beneficial for the actor in the short term. The Panopticon design shapes an ideal world where everything and everyone is always visible and where one always immediately experiences the right consequences of one's deeds. As a result, people will always act in accordance with the rational moral principle of utility for maximizing happiness.

Dystopian Technology

Such a positive attitude toward technology is not universal. Developers in the workshop remarked that they do see that there are always people who do not actually want so much technical innovation at all. Moreover the utopian idea of perfect convergence of smart technology with the demands of people can turn from a *Utopia* into a *Dystopia* when as a result people never ever would have to leave their house anymore. This was expressed by a workshop participant with reference to the movie *WallE*, where robots do all the work, but this has rendered people fat, immobile couch potatoes. Another *dystopian* danger mentioned was that all sensors and databases might fall in the hands of a totalitarian regime.

Dystopian technology can be explained as *the accumulation of technology into a system that takes control of humanity, with the complementary ethical calling to put limits to this rush of technology*. Following a fully dystopian view, technology in itself is principally dangerous. The totalitarian exploitation of technologies would not be due to wrong use, but it is a pattern residing in technology itself that when utopian ideals become realized, they turn out dystopian.

The dystopian view of technology combines with duty ethics rather than with utilitarianism. In the modernistic search for an ultimate rational moral principle, the theory of duty ethics by Immanuel Kant (1724–1804) is the competitor of utilitarianism. According to Kant the actor's intention and not the effects of an action count for ethical evaluation. Actions are qualified ethically good if they are based on good intentions, meaning that they stem from duty, from respect for the moral law. Kant stresses that this assumes the possibility of a free will to determine one's own actions: autonomy. Unlike Bentham, Kant did not explicitly consider the impact of technology, but the emphasis on human autonomy can be seen as the ethical complement of the dystopian view of technology. The reversal from utopian to dystopian technology can be illustrated with the adventures of Bentham's Panopticon. At the end of the twentieth century, since the famous analysis by Michel Foucault (1977), the Panopticon has rather become an emblem of a *Dystopia* of social control by technology. Foucault does, however, not himself endorse duty ethics, but for example the call for an "imperative of responsibility" by Hans Jonas (1984) is an explicit expansion of Kant's duty ethics to technology.

Ambivalent Technology

The overall atmosphere among the participants in the interactive screens workshop could rather be called *ambivalent*. The focus was on the need to find balance between the positive opportunities of technology and the negative impacts and risks. It was felt that surveillance may be presented too easily as dystopian (with reference to Big Brother or the Panopticon), whereas it is undeniable that surveillance does also help to actually prevent assaults. Somebody recognized that one's evaluation of technology is linked together with one's theoretical understanding of technology, and even referred to Actor-Network Theory (cf. Latour 1994), with its emphasis on the intertwinement of humans and technology. Other remarks that expressed an *ambivalent view* were that it is a matter of finding balance, of finding or creating possibilities to turn technology off again, or of tweaking technology; education is important; people should have a choice and they need to become proficient to recognize and make choices. Regulation was mentioned as a remedy for making the behind-the-back systems more transparent, for example a certification register for sensors. An idea for design improvement was the notice board as a model for the interaction with the screen, which would give the public a more active role in the interaction compared to the sensing and automatic content collection system, which is top down controlled.

In the Product Impact Tool ambivalent technology is the view that acknowledges that *human existence is unescapably mediated by technology with always both good and bad effects, prompting an equally complex ethical challenge, which is to cope with technology in a balanced way*. This means a middle position between the utopian and dystopian views, which does not mean an easy solution, and it does not diminish the importance of technology. "Technology is neither good nor bad; nor is it neutral" is the nice formulation (by Kranzberg), emphasized in the ethics of technology by Michel Puech (2016, 2).

With the effects of technology ambivalent and the focus on coping in a balanced way, this view of technology combines better with practical approaches in ethics which we see in the postmodern revival of ancient focus on virtues and the care of the self. Balance has since the time of Plato and Aristotle always been a key feature of virtues. Michel Foucault's work to revive the ancient care of the self (Foucault 2000a) allows to rethink human freedom and autonomy as the practice of coping with external influences, instead of the opposite of determination (in line with Kant's ethics). By the way, Kant himself does explicitly acknowledge that the existence of free will is a philosophical enigma and impossible to explain in the scientific framework of physical and social determinations (nor of technical conditions we may add). Foucault thinks that this trait in Kant marks the inauguration of modern philosophy in its true sense. Here Foucault does not so much think of Kant's ethics of free will and duty, but especially of Kant's more pragmatic work on anthropology and his essay on the Enlightenment. Foucault finds in Kant the beginning of a critical awareness of the paradoxes and side effects of progress, whereas classical rationalism before Kant was naïve and too optimistic. In his late work, Foucault was comparing and combining the critical "attitude of modernity" (Foucault 2000b) with the theme of the "care of the self" from ancient ethics (Foucault 2000a). This combination is inspiring for a practical ethics for coping with the impact of technology, and finding a good balance of humans and technology.

Conclusions

The workshop with the Product Impact Tool about interactive screens research within the research project OBSERVE was used to show how an impact assessment with this tool did evoke reflection and discussion about ethical concerns. Many points from the discussion have been reported on here, and this gives an idea of how in the practice of a workshop the movement from impact assessment to ethical concern occurs all the time.

A further question in this chapter was how the connection between impact assessment and ethical concern can be understood in a more conceptual way. After a reflection on the conceptual affinity and reciprocity between figures of impact and figures of ethical concern, I have reviewed all twelve types of impact and discussed which ethical concerns typically are raised in connection with any of them. It appeared that some types of impact have predominantly a negative ethical value (a critical concern) and others a positive (an ethical remedy). In the positive form an impact figure can be used as a strategy for responsible design that remedies some ethical concern. Along this structure the following table summarizes the output of the workshop and further elaboration.

Type of impact	Ethical concern/negative value	Ethical remedy/positive value/design strategy
Utopian technology	Only scarcity and unfair distribution	Utilitarian ethical principle; Human completion, optimization
Dystopian technology	Technology takes command	Deontological ethical principle; Limits to technology, precautionary principle
Ambivalent technology	Technology neither good nor bad nor neutral; Ethics of virtues and arts of living	Practical ethics of virtues and arts of living; Hybridization, balance, re-humanization
Guidance	Misguidance, nonguidance	User-friendly; manuals, instruction
Persuasion	Who controls?; unawareness, wrong reasons/deception	Behavior support
Image	Stigmatization; superficiality	Positive association, growth, self-esteem
Coercion	Interference with freedom and responsibility	Can't go wrong, one way of using
Embodiment	Slippery slope, bad habits, addiction	Natural self-extension, user-friendly
Subliminal affect	Subconscious drives, temptation, unawareness	Positive sensory stimuly; a comfortable, welcoming, healing ambiance
Side effects	Unintended consequences, collateral damage, risks	Anticipation, impact assessment; win-win situation
Background conditions	Dependence; withdrawal and unawareness	Raising awareness; Integration, system design, regime change, revolution
Technical determinism	Negative disruption, shifting values and preferences, decay	Positive disruption; resistance, subversive use of technology

A few points of discussion may be added to these results. This exploration of impacts and ethics is based in practice and connected to the case of interactive screens. Still, the results have wider application in the case of other interactive technologies and technology in general. In the same way as the twelve types of impact in the tool, the overview provides a framework, scheme, vocabulary, which is helpful and necessary for interpreting and articulating the adventures of the interactions between humans and technology. It remains also provisional and is only one possible framework. The repertoire of impacts and concerns helps to dive into an exploration of the details of the interwovenness of humans and technologies, of our own technically mediated existence.

The Product Impact Tool does contribute to the ethics of technology in a certain way. It does not provide a clear answer or a method to decide in case of an ethical dilemma. The last step of making a decision for action is left to the wisdom and responsibility of those engaged with the issue. What this tool does do, in case of a given dilemma, but just as well if there are no known issues yet, is that it contributes to awareness and insight about the impact of technology, and it does evoke reflection and discussion about ethical concerns. The Product Impact Tool therefore does stimulate a responsible attitude of engineers, designers, and other stakeholders in innovation.

References

Dorrestijn, Steven. 2012. *The Design of Our Own Lives: Technical Mediation and Subjectivation After Foucault* (PhD thesis). Enschede: University of Twente. http://www.stevendorrestijn.nl/downloads/Dorrestijn_Design_our_own_lives.pdf.

Dorrestijn, Steven. 2016. "History, Philosophy, and Actuality of the Utopian View of Technology: On Pierre Musso's Critique of Network Ideology." In *Pierre Musso and the Network Society*, edited by José Luís Garcia, 103–129. Cham, CH: Springer International Publishing.

Dorrestijn, Steven. 2017a. "The Product Impact Tool: The Case of the Dutch Public Transport Chip Card." In *Design for Behaviour Change: Theories and Practices of Designing for Change*, edited by K. Niedderer, S. Clune, and G. Ludden, 26–39. Abingdon and New York: Routledge.

Dorrestijn, Steven. 2017b. "The Care of Our Hybrid Selves: Ethics in Times of Technical Mediation." *Foundations of Science* 22 (2): 311–321.

Eggink, Wouter and Steven Dorrestijn. 2018. "Philosophy of Technology x Design: The Practical Turn." In *Proceedings of DRS 2018: Design as a Catalyst for Change*, edited by C. Storni, et.al., Vol. I, 190–200. London: Design Research Society. http://www.stevendorrestijn.nl/downloads/Eggink_Dorrestijn_Philosophy_x_Design.pdf

Fogg, B. J. 2003. *Persuasive Technology: Using Computers to Change What We Think and Do*. Amsterdam and Boston: Morgan Kaufmann Publishers.

Foucault, Michel. 1977. *Discipline and Punish. The Birth of the Prison*. New York: Random House.

Foucault, Michel. 2000a. "The Ethics of the Concern for Self as a Practice of Freedom." In *Ethics. Subjectivity and Truth: Essential Works of Foucault 1954-1984. Vol. I*, edited by Paul Rabinow, 303–319, London: Penguin.

Foucault, Michel. 2000b. "What Is Enlightenment?" In *Ethics. Subjectivity and Truth: Essential Works of Foucault 1954-1984. Vol. I*, edited by Paul Rabinow, 281–302. London: Penguin.

Ihde, Don. 1990. *Technology and the Lifeworld: From Garden to Earth*. Bloomington: Indiana University Press.

Jonas, Hans. 1984. *The Imperative of Responsibility: In Search of an Ethics for the Technological Age*. Chicago: University of Chicago Press.
Kudina, Olya and Peter-Paul Verbeek. 2019. "Ethics from Within: Google Glass, the Collingridge Dilemma, and the Mediated Value of Privacy." *Science, Technology, & Human Values* 44 (2): 291–314.
Latour, B. 1994. "On Technical Mediation." *Common Knowledge* 3 (2): 29–64.
McLuhan, M. 2003. *Understanding Media: The Extensions of Man* (Critical edition by W.T. Gordon). Corte Madera, CA: Gingko Press.
Norman, Donald, A. 1988. *The Psychology of Everyday Things*. New York: Basic Books.
Puech, Michel. 2016. *The Ethics of Ordinary Technology*. New York: Routledge.
Swierstra, T., D. Stemerding, and M. Boenink. 2009. "Exploring Techno-Moral Change: The Case of the Obesitypill." In *Evaluating New Technologies*, edited by P. Sollie, and M. Düwell, 119–138. Dordrecht: Springer.
Tenner, E. 2003. *Our Own Devices: The Past and Future of Body Technology*. New York: Alfred A. Knopf.
Thaler, R.H. and C.R. Sunstein. 2008. *Nudge: Improving Decisions About Health, Wealth, and Happiness*. New Haven: Yale University Press.
Verbeek, Peter-Paul. 2005. *What Things Do: Philosophical Reflections on Technology, Agency, and Design*. Pennsylvania: Pennsylvania State University Press.

9 POSTPHENOMENOLOGY OF AUGMENTED REALITY

Galit Wellner

Introduction

The summer of 2016 could be marked as the summer of Pokémon Go, the augmented reality (AR) game with which people—children and adults—spent hours and days chasing after imaginary creatures in real places. Simply explained, the technology requires people to point their cellphone's cameras on spots around them and then it shows that location on the screen of the cellphone with a Pokémon on it (or two if you are lucky!). The display is in real time and interactive, so you can immediately see if you can catch one: "have to catch them all" is the game's slogan.

AR existed well before Pokémon Go. This technology has been implemented in various applications, such as: selling furniture or clothes where the display simulates the look (of the person or of the room) after the potential purchase; professional on-the-job training demonstrating how a certain action should be performed on the background of the real workshop, machine, and so on; or translations of street signs and other texts for tourists traveling to foreign countries.[1]

Lev Manovich (2006) defines AR as "the layering of dynamic and context-specific information over the visual field of a user" (p. 222). At the core of his definition lies the concept of layers of information, which are shown on top of the actual reality. The information can be "real" in the form of products' prices, people's names, or the names of star layouts up in the sky; and it can be imaginary, such as Pokémon graphics or personal assistants, whose voice may lead one to think a real person is out there. The last example of personal assistants reveals that Manovich's definition can be expanded beyond the visual into the auditory.[2] For both the auditory and the visual, the display of layers of information leads to the emergence of a new space, termed by Manovich "augmented space" to denote "the physical space which is 'data-dense', as every point now potentially contains various information which is being delivered to it from elsewhere" (p. 223).[3]

Another element in Manovich's definition is the context in which the information is displayed or heard. The information should be location sensitive as well as relevant to the time of its actual display. An additional element is personalization that filters the information according to the identity of the user, her history, and her preferences. Hanna Schraffenberger and Edwin van der Heide (2013) explain that merely adding information is not enough and there should be correlations between the virtual and the real. Thanks to these correlations, reality is augmented, rather than just added with information. Beyond the observation made by Schraffenberger and van der Heide, it is important to notice that personalization and contextualization not only contribute to the AR experience but also assist in accommodating vast amounts of information to a limited display space (or time frame in case of audio augmentation).

Nicola Liberati and Shoji Nagataki (2015) offer an updated description of AR, according to which the layers can include—in addition to information—augmented objects. They argue that while the "standard" information element of AR focuses on the user and attempts to improve her perceptions, the augmented objects modify the world: "The world becomes 'embedded' with augmented objects as well as normal objects" (p. 135). This situation is also known as "mixed realities" (Liberati 2017, 3) because the world contains a mixture of real and augmented objects. The Pokémons are a good example of augmented objects.

In order to explore the AR experience, I suggest studying the relations between AR technologies, their users, and the world as shaped by the technology, or in postphenomenological terminology—what world is mediated for us by AR technologies and how (see Ihde 1990; Verbeek 2005). Postphenomenology is a branch of philosophy of technology originally developed by Don Ihde, which explores our relations with technologies and the world. While classical phenomenology seeks to systematically analyze our experience of the world, postphenomenology goes one step further. Realizing that today fewer and fewer aspects of the world are unmediated by some kind of technology, it studies the relations between three constituents: humans, technologies, and the environment. They are presented as a scheme: I-technology-world (Ihde 1990).

I shall examine the postphenomenological relations involving AR technologies in three steps. The first follows the classical postphenomenological relations and asks what does it mean to add layers on reality. The second step focuses on Peter-Paul Verbeek's notion of composite intentionality and questions whether AR technology is just adding layers to the representation of the world or modifying the experience. The third and last step explores what happens when augmented objects demonstrate their own intentionality—that is, when the information layers become more dominant than reality itself.

First Step: Classical Postphenomenology

Whereas Martin Heidegger's tool analysis identifies a dichotomist set of relations between humans and technologies—of presence at hand versus readiness to hand—postphenomenology accentuates the multiplicity of types of relations we develop with our surrounding technologies and the various ways in which these technologies mediate the world for us (Ihde 2010). Postphenomenology represents this variety by permutations of its basic scheme of I-technology-world, using parentheses and arrows.

The first type of relations is *embodiment relations*, in which we experience a given technology as part of what Maurice Merleau-Ponty names our body scheme. When I drive my car, I experience the world as if my body and the car's body are a single unit. When I park the car, I know how much I can drive backward. And when I drive in narrow streets, I "sense" where the combo of me and my car can pass. This type of relations conforms very well with Heidegger's notion of readiness to hand, in which the technology withdraws to the background and the user concentrates on the work to be done, like driving. The postphenomenological scheme looks like this:

$$(I - technology) \rightarrow world$$

AR technologies must withdraw to the background in order to allow the user to experience the layers on top of the reality. This characteristic is also shared by virtual reality (VR), and by the same logic. The technological artifact must become what postphenomenologists term "transparent" (e.g., Van Den Eede 2011). For both AR and VR, the technological device is experienced as part of the user's body.

Another type of postphenomenological relations is *hermeneutic relations*, in which I experience the world and the technology as a single unit. Put differently, while in embodiment relations "I" and "technology" are experienced as a combined unit, here the technology and the world function as a combo. When a technology is experienced as part of the environment, it tells us something about that environment. These relations often involve media that is read and deciphered, and hence the term "hermeneutic relations." Think of a watch. We can have embodiment relations with it and feel it as part of our hand, but this description does not reveal much about the meaning of the watch. We need to refer to the "text" it displays, that is, the hour, whether displayed in analog or digital form. The "text" is understood as part of the world that is "out there" waiting for us to become meaningful. The postphenomenological scheme represents this relation as:

$$I \rightarrow (technology - world)$$

AR technologies offer a unique form of hermeneutic relations in which they attempt to show a unified experience of media and world. The information is not just information about the world, it is part of the world. Here again it is interesting to compare to VR, in which media attempts to *replace* the world. In AR, the world remains as it is, but it is augmented by the information.

The next postphenomenological relation is *alterity relations*, in which the technology is experienced as a quasi-other. Classic examples are children's dolls and ATMs, with which we maintain a certain dialogue. In these relations, the world withdraws to the background, and so in the formula, the world is within the parenthesis:

$$I \rightarrow \text{technology } (-\text{ world})$$

Some layers of AR are experienced as alterity relations. My opening example of Pokémon Go fits within this category, where people relate to Pokémons as entities with which one can communicate (they don't have to answer…). Another example is personal assistants like Siri, Cortana, and Alexa, which are designed to maintain alterity relations with their users.

The fourth type of postphenomenological relations is *background relations*, where the technology operates in the background, unnoticed, like air-conditioning and lights in a room. While in alterity relations the world is in parentheses in order to show that it occupies the background, here the technology is in parentheses for the same reason. The postphenomenological scheme is:

$$I \rightarrow (\text{technology } -) \text{ world}$$

When technological devices operate properly, as in the case of well-functioning AR hardware, we don't pay attention to the technology and just experience the world. But the world is not experienced "as is," as if the technology does not exist. Just like the air-conditioning that leaves the room cool, AR technologies leave traces that are noticeable (see Wellner 2017). Unlike embodiment relations in which the technology withdraws to the background but becomes part of our body scheme, here the technology withdraws to the background while becoming part of the world.

Each of the four relations covers a certain aspect of the AR experience, in line with older technologies, like cars and ATMs. But AR technologies offer a new experience, and this newness remains latent. In the next section, I will review a new type of relation that may provide a more comprehensive description of the experience AR technologies offer us.

Second Step: Composite Intentionality

The four basic postphenomenological relations were originally formulated by Ihde and then expanded by Verbeek (2008a). Verbeek's extension may allow us to understand the special relations people have with their AR technologies. He starts by elaborating on the notion of "technological intentionality" to model the tight relations between humans and contemporary technologies, explaining that "'technological intentionality' here needs to be understood as the specific ways in which specific technologies can be directed at specific aspects of reality" (2008a, 392). Technological intentionality is interpreted as the ability of technologies to form intentions so that they enable the users to do things which could hardly be done without such technologies. Verbeek explains how technological intentionality operates: "Even though artifacts evidently cannot form intentions entirely on their own… because of their lack of consciousness, their mediating roles cannot be entirely reduced to the intentions of their designers and users either" (Verbeek 2008b, 95). For Verbeek technological intentionality supports human intentionality so that "when mediating the relations between humans and reality, artifacts help to constitute both the objects in reality that are experienced or acted upon and the subjects that are experiencing and acting" (Verbeek 2008b, 95). In this step I will show how AR technologies practically shape reality. I will refer to Navigation apps in order to demonstrate how technological intentionality works. These apps direct the user to go through certain streets, may they be toll highways or the narrow streets of a quiet neighborhood, to mention two of the common "accusations" against these apps. This kind of apps reveals a double meaning of intentionality—on the one hand as directing the way and on the other as "channeling" of a meaning.

Based on the notion of technological intentionality, Verbeek extends the postphenomenological relations in several directions; one of them is termed "composite intentionality." It indicates situations in which not only human beings but also the technological artifacts they are using have intentionality. In composite intentionality, the "directedness" of a technology is added to the human intentionality. The intentionality is added hermeneutically; that is, it adds new ways in which the technology "reads" the world. This structure of double intentionality is represented in the postphenomenological scheme with two arrows:

$$I \to (technology \to world)$$

This is a permutation of the hermeneutic relations' formula in which the technology and the world are connected by a dash. The arrow indicates intentionality, and within the parentheses, it is no longer associated solely to the human "I." The technology is imbued with some independence and ability to decide, to take direction (see also Wellner 2013).

When Verbeek published his article in 2008, many of his examples were speculative and taken from the field of art. Today, we can demonstrate his theoretical developments with everyday AR technologies. Let's take a basic AR situation: driving using a navigation app like Waze. The app provides several layers of information on top of the reality of roads represented by a map (we take it for granted that the map representation is accurate and fit the world of roads and traffic signs). When projection technologies mature, it might be possible to display the layers on the windshield (see Michelfelder 2014). That will enable "full AR," in which the reality—as seen through the glass—would serve as the basic layer on top of which the information layers would be displayed. But today the basic layer is a digital map. In both versions of a map on screen and reality through the windshield, the layers are hermeneutic by their nature because they show information that is perceived by the users as "the world out there" and that requires reading and deciphering.

Navigation apps' elementary layer consists of directions of how to get from point A to point B. Some apps offer a second layer with the dynamics of traffic— flowing, moving slowly, stuck. And when stuck, the apps can calculate how long you are likely to remain stuck and update the estimated time of arrival or offer an alternative route. There is a third layer of advertisements that recommend to the driver (or the navigator) businesses and activities on the way. The directions and suggestions may change the original route so that the driver's intentionality is not as "pure" as driving without the app and its recommendations. The intentionality is "composite," combined of the user's human intentionality and the software's technological intentionality (see Wiltse in this book).

Sometimes the technological intentionality of the algorithm takes over, as the following example demonstrates: Some years ago, winter rains were more intensive than the usual and Tel Aviv's Ayalon highway was flooded. The police closed all entries to the highway and announced the closure on the mass media—radio, television, newspapers, and even major websites. But apps, back then, were not considered media; they were taken instrumentally as nice gadgets that help drivers to navigate. Waze's algorithm analyzed the traffic and realized that the Ayalon highway was empty, while all the surrounding roads were jammed. It immediately directed its users to the highway, thereby contributing to even larger traffic jams. After that incident, the software developers connected the app to police announcements, and today the app takes into account not only current traffic conditions but also planned and unplanned road blocks. In this story, the world—in the form of roads and traffic—and technology—in the form of cars and navigation apps—cannot be taken as passive actors who just respond to active humans. The decisions that drivers are taking cannot be understood with the classical tools of "subjectivity-objectivity," "free will," or "autonomy."

Let's examine the function of the left arrow in the scheme, that which connects the human user to the combo "technology-world." This arrow represents the human intentionality. The arrow can be "projective" when it represents "a focused reference to the world" or can be "reflective" when referring to "a movement from the world" (Ihde 1979, 27). Projective intentionality plays a major role when the user chooses a route different from that recommended by the app. This is what I do when I drive with Waze: I purposefully maintain my focused intentionality and do not automatically accept the app's instructions. First, I ask for alternative routes to select the one I feel more comfortable with, like refraining from toll roads or preferring a scenic road. My choice is often different from the app's recommendation, which is usually the fastest route, even if it is much longer and even if it is more prone to traffic jams. Sometimes, the difference between the recommended route and the second one is only few minutes. After choosing a route, the app may alert that a traffic jam is developing somewhere down the road. The app is likely to suggest an alternative route that bypasses that traffic jam, and it will blip when suggesting so. It is up to me to "obey" and follow this new route or choose another one. Here, the technological intentionality does not preclude my own human intentionality. They both coexist and co-shape each other in an endless loop.

These interactions between me and the app have an effect in and on the world. Occasionally I take the app's advice and turn into a small quiet street hoping to arrive on time at my destination. But then I see more drivers going before and after me, taking the very same turns, apparently many of them are users of the app. Eventually, we all form a new traffic congestion, albeit alleviating the jam we all tried to bypass.

The double intentionality as formulated by Verbeek's composite intentionality well represents the relations with current AR technologies that do not merely add information but augment a specific place and time while interacting with the user. Here, more than ever, users matter. Time and again we realize that people, technologies, and world co-constitute each other and remain interdependent. Navigation apps shape driver's perception on traffic jams, co-determine the way in which the driver arrives at her destination, and co-shape the traffic world by directing more drivers to a certain route.

Composite intentionality may explain how AR technologies affect human behavior. However, the scheme representing composite intentionality assumes a certain flow going from the human to the technological and the world. It hardly fits situations in which we "give up" our intentionality; delegate our decision making to a technology; and let the technology lead, decide, and operate. A simple example of this experience can be that of Waze's users who delegate the determination of direction to the app without thinking of the route. They drive according to the app's instructions, even when erroneous. Once in a while they find themselves going in loops when the app cannot determine the way or has other difficulties. Composite intentionality cannot accommodate situations in which the human intentionality "withdraws." A new type of relations is required.

Third Step: From Intentionality to Relegation

In the third step, I examine the effect of reversing the human intentionality arrow in the postphenomenological scheme so that it points *to* the human and not *from* the human:

$$I \rightarrow (\text{technology} \rightarrow \text{world})$$

In this permutation, the human intentionality "withdraws," and the technological intentionality "takes over." It reflects situations in which technologies control the world as well as the users. Drivers that go in loops because they obey the navigation app's instructions are just the harbinger. The more AR technologies become part of our everydayness, the more examples will emerge. Keiichi Matsuda lays out a striking look into such a future in his artistic video work titled "Hyper-Reality" (https://vimeo.com/166807261). While the Ayalon Highway example provided a glimpse into an AR-intensive future, the "Hyper Reality" clip can be read as an extensive speculation about the human experience in such situations. Six minutes long, the clip shows a possible reality where public transportation, streets, and supermarkets are digitally decorated with colors, badges, and augmented objects, and every move is accompanied by an overwhelming amount of information. The clip shows the shift from mediation to an extremely active shaping of reality and human intentionality that tends to amount to a form of control. In this reality, the division between users' and technologies' actions (or between trace and substrate that responds to an activity (Wiltse 2014)) blurs. Who acts? Who controls? With intelligent layers and sophisticated augmented objects, the answers become more complex.

Matsuda's clip can be divided into three parts: the first shows an everyday situation of commuting by public transportation where AR technology shapes every minute and every centimeter; in the second part, the protagonist experiences an attempt to hack her digital identity; and in the third part, she ends up selecting a new identity. The clip occurs in three different locations: a bus, a supermarket, and a street. Each place exemplifies the visual transformations AR brings, sometimes in a sharp contrast to a gray and dull reality. The price for this augmentation is an inability to look, to gaze, to decide. It is evident from the first minutes but becomes acute toward the end of the clip.

Clip—Part I:

In the first minute of the clip, Juliana the protagonist sits in an augmentedly colorful bus, playing a mobile game that occupies most of her visual field and exceeds what we know today as the cellphone's screen. For a moment she wonders about her life and career, but changing her life requires that she reset her virtual identity and lose the points accumulated for that identity. It seems like too high a price, and she declines. In the next scene, we see her shopping in a supermarket digitally augmented with information and multicolor decorations.

All the figures with which she communicates in the clip—her personal assistant (named "inspiration guru"), the shopping assistant, and the customer support personnel—are avatars, purely virtual. These chatbots function as personalized AI layers so that, for example, the avatar for customer support looks like an elder brother who gives good advice, and the shopping assistant is a cute little puppy that recommends products to buy as if it is a kid asking a parent to buy a candy. The personalized chatbots reinforce a certain identity on the protagonist, as a passive woman with no aspirations. This reinforcement becomes almost brutal when her "inspiration guru" scolds her—"you are late" (0:25)—and refuses to help her find another job as a teacher, saying, "Trust the app; it always chooses the right jobs for you!" (0:41). It is a manifestation of a technological intentionality that "takes over" the human intentionality. As noted above, the notion of technological intentionality refers to artifacts' capacity to "help to constitute both the objects in reality that are experienced or acted upon and the subjects that are experiencing and acting" (Verbeek 2008b, 95). In the clip, the artifacts do much more than merely "helping" to constitute a reality. The algorithms control reality itself as well as their users—Juliana's career, wishes, and desires.

It is an extreme delegation, compared to Bruno Latour's door example (Latour 1992), where the human groom was replaced by a door technology. In Latour's example, closing the door is "delegated" to an automatic groom. Here, however, the human remains in the picture, but only as a passive participant. I term this situation "relegation" (thanks to Dylan Wittkower for offering this word!) because the human intentionality is assigned to an inferior position. The algorithms "downgrade" the human intentionality. In other words, not only there is no symmetry, control now resides on the side of the nonhumans.

Once the AR layers gain autonomy and intelligence, they shape the real, not only in the short term of a traffic jam (as the Waze example shows) but also for the long term in the form of career selection. The augmented world is not passive, nor is it just (digitally) responsive. Rather, it is proactive and it shapes users' perceptions, actions, and desires. Unlike contemporary AR games, like Pokémon Go that create a fictional world in parallel to the real, here the reality and the game are intertwined. Reality becomes mixed with the game, combining the real and the augmented in a way that a demarcation between reality and fiction becomes near-to-impossible. No less simple or easy is leaving the augmented space, because of the addictive character of the personalization and gamification. The chatbots, the games, and the advertisements form a world in which users-players become "slaves of a single reality" (Liberati 2017, 16).

Clip—Part II:

While in the first part of the clip the technologies seemed to focus on the protagonist (albeit not always on her preferences), in the second part her illusion of being the center of interest is cracked once her identity is hacked. This moment occurs when she is in the supermarket, examining a yogurt. The texts in the augmenting layers promote the product as a means of losing weight, as healthy, and even as something that can make her beautiful. All of a sudden, the yogurt's front declares "for real men only" and the background badges proclaim "Fitness fuel, build muscle, increase fertility" (2:55). At the same time, her shopping assistant switches from a sweet puppy into a "sexy girl" avatar. These changes are the sure signs of hacking as they reflect the preferences of a male hacker. The protagonist contacts the customer support center, but instead of the "elder brother"-like chatbot, she receives a sexy assistant addressing her as "Emilio" (although her name is Juliana).

The hacking can be conceived as a breakdown that makes the AR tool noticeable, what Heidegger terms present-at-hand. The AR software should be hacked in order to come to the front and gain the user's attention. But the Heideggerian terminology is not sufficient because what matters here is not the ability to notice the malfunctioning technology but rather the loss of control, which is felt when the technology does not operate properly. In the beginning of the video clip, the protagonist had a feeling of control and centrality. Once hacked, these turn out to be just an illusion, provided and maintained by the AR software. The video clip reminds us that we are not at the center of the world. Apparently, thinking that everything is out there "for me" requires an extensive technological layer.

The protagonist's response to the attempt of hacking looks reasonable: she follows the customer support's advice and restarts her device. Suddenly, all the badges, advertisements, and music disappear, and the aisle of the supermarket looks gray and dull. In the background we can hear a baby crying. When the system is up again, the customer support avatar returns as the brother figure and tries to communicate a "business as usual" atmosphere by telling her: "You look fantastic today. New shoes?" (min 4:20). Obviously the avatar cannot see her visually, and these words are a kind of small talk aiming to please her. At this stage she is still worried about her points, and the chatbot instructs her to verify her biometric information at the service center.

Clip—Part III:

The protagonist is out in the street, guided by the AR app. There is a blue line virtually marked on the sidewalk directing her to the service center. The buildings and the sidewalks are colored, and augmented objects like trees and flowers are added to embellish her way. The people who walk on the street toward her are accompanied by virtual badges designating, for instance, where to buy a shirt "like this." What seems to be a routine AR walk is disrupted by a shadow quickly approaching her. This figure cannot be identified and has no badge; it is marked as blinking and blurring. Within a second or so, the figure reaches her, injures her palm, and runs away. The AR software immediately marks her hand by a badge "injury detected" (5:03), suggests she follows a red line marked on the floor, and asks if she wants to reset her identity. The app fails to classify the event as a biological hacking, probably because it can assign meaning only to events within the augmented space and according to its rules.

The biological hackers took all the points, and Juliana's virtual identity has been deleted. And whereas she feels pain in her hand, what really bothers her is the loss of the points and she mumbles, "The points! The points" (5:16). No doubt she is emotionally attached to her virtual points, and when they are gone, it is a moment of grief (for attachment to things see Puech (2016) and Puech in this volume).

Meanwhile, the augmented space fades away. In the place where until a second ago was a huge provocative advertisement for plastic surgeries, now we see a humble statue of Jesus and above it a colorless sign stating, "A New Life Has Begun!" (5:23). By the time Juliana crosses the road and reaches the statute, the software has restarted, and it offers her a new identity—a religious one. She eventually manages to follow her own intentions only when the system breaks following the hacking.

If in the first minutes of the clip the protagonist considered for a moment the option of resetting her AR identity (just to omit it and return to her "normal" augmented environment), toward the end of the clip she is forced to do so.[4] In the cosmos of "Hyper-Reality," resetting the virtual identity means returning to reality (cf. Liberati 2017), but the protagonist hesitates to do so because the "naked" reality is depressingly gray and derelict.

But being gray and depressing is not equal to passivity. The hacking reveals a world that is active and that refuses to hide behind the colorful augmented objects and badges. Alternatively, the hacking can be regarded as an "invasion" of the reality into the augmented space. Although violent and daunting, the hacking gives the protagonist an opportunity to change. The hacking is indispensable because under the rules of the augmented space, such a change is undesirable and practically impossible. Playing according to different rules is equal to resisting the system. This is the essence of being subversive. The system has no tools to deal with such actions.

The emergence of the blurring figure of the hacker accentuates the wild aspects of the world that exceed the "I-technology" relations. An attempt to describe the hacking scene in postphenomenological terms leads me to form a new permutation in which the world has its own intentionality. This intentionality is separate from that of the protagonist, whether a well-functioning AR technology dictates a certain worldview or a hacker disrupts the proper functioning of the system. The new permutation of the scheme reverses the right arrow:

$$I \rightarrow (technology \rightarrow world)$$

In the previous section I discussed the reversal of the arrow that connects technology and the world.[5] Now, when both arrows are reversed, a new meaning of technological intentionality emerges. Matsuda's clip offers an extreme form of technological intentionality, which "forces" the user to obey the technology. Here the unintended consequences (in the form of the hacking) come from the "world" component, not from the technology or the users. It is a new kind of technological intentionality that is assigned to the "world" at the expense of the users and their technologies.

Such an interpretation of technological intentionality leads to a new understanding of the human intentionality, which can be analyzed in terms of gaze or view. In the clip the protagonist has a limited control of her gaze as she sees only what the AR software presents to her and cannot see what is covered by layers of fictional information. The AR system hides parts of the real with games, advertisements, and augmented objects. The protagonist could not notice the statue of Jesus until the AR device shuts off. The statute was hidden behind an advertisement. There was a very short period of time (when the AR system restarted), in which she could see reality as is and reconstruct a new identity. We can wonder what could have happened had she seen another real object or had more time to think freely.

As long as the AR system operates, every moment in her life is mediated and her attention is led and managed by the AR algorithms. William Uricchio identifies this phenomenon as paradigmatic to AR technologies: "Through the lens of the AR-activated device, there is no such thing as an innocent gaze: the act of gazing and the views consequently seen are transformed into a process of signification as images are laden with particular meanings… The innocence of uninformed exploration… is transformed into an encounter with the always-already-marked and significant" (Uricchio 2011, 32). In the clip (as in most algorithms today), the parameters for significance are not set by the user but rather by an algorithm that the user does not control. It is interesting to realize that the clip is narrated through the gaze of the protagonist, and we as viewers do not see her. We see only what she sees through the AR system. Even our gaze as viewers is controlled.

Is there a way to escape our AR future? Hyperreality's solution looks familiar. Like Heidegger's famous statement in his last interview (Der Spiegel, May 31, 1976), "only a god can save us now," here the Christian religion pops up and offers salvation to the wounded body (and soul) of the protagonist. A secular interpretation of the final scene of the clip may accentuate the return of the body. Once the protagonist is injured, the illusion of the AR world evaporates and reality comes back, for better and for worse. My own lesson is realizing there is a need to develop digital skills that will enable us to understand when our control is taken by algorithms or the world and find alternative behaviors to restore our control of our fate.

Conclusions

This chapter was composed of three steps, each accompanied by an example, arranged by the "immersiveness" of the AR technology and degree of necessity: going from the AR game Pokémon Go, through the context-specific navigation app Waze, to a speculative video clip showing an imaginary full AR app that explores what life would look like if and when AR technologies become interwoven into our everydayness.

The steps were accompanied by postphenomenological analyses developing from the 1980s and the 1990s, through the 2000s to new options. The first step contained an overview of the postphenomenological basic relations—embodiment, hermeneutics, alterity, and background. In the second step I described composite intentionality, a relatively new type of postphenomenological relation in which intentionality is not limited to humans but can also be imputed to technologies. In the third step I "switched vehicles" and used an exploratory example of AR in order to construct an extreme form of technological intentionality termed "relegation." The situations depicted in Matsuda's clip are more than "algorithmic interventions between the viewing subject and the object viewed" (Uricchio 2011, 25). They shape the user's gaze, behavior, thoughts, and desires. The third step is an attempt to understand how AR technologies can become part of our views, actions, and thinking. The new term "relegation" reflects such a complexity.

Where is the human in the future landscape of AR and AI technologies? Uricchio suggests that we are no longer "world builders," as Heidegger framed (2011, 33). The world is no longer a picture. We are now "in" the world like parts of a huge machine controlled by algorithms. This is also what Shoshana Zuboff (2019) revealed when she studied the "economies of action" of Facebook and Pokémon Go. She describes how these companies treat their users as data generators that can and should be manipulated through "behavioral modifications" without their knowledge or consent. Thus, intentionality can be imputed to technologies and the world, and the notion of relegation becomes central to the relational analysis.

While usually most analyses focus on the human-technology micro-cosmos, contemporary technologies like AR demonstrate that it is not sufficient to examine how humans and things interact. There is a need to take into consideration the world. It becomes acute with AR technologies that have a special relation to the world because they do not just mediate a given world that is "out there," neutral and unchangeable. They change the world in various ways. The postphenomenological scheme already includes the world, thereby reminding us to make it part of the analysis of human-technology relations. However, there is a need to further develop the role the world plays, beyond the micro-cosmos of I-technology. More importantly, it may be an opportunity to raise awareness to the problematics of the Anthropocene and lead people to rethink the world in the context of the technologies they use and their responsibility for the changes in it.

Notes

1. AR technology requires specific hardware and software; yet, the examples mentioned were mostly software-based, using the cellphone as their hardware (see Wellner 2015). Attempts to develop special hardware are still underway. For example, the Google Glass project originally aimed to provide "true AR," though it did not deliver the promises made earlier in the project's inception phase. The Google Glass project might be history by now, but cellphone-based AR is live-and-kicking as the Pokémon Go craze proves.
2. For example, the app "Zombies, Run!" simulates imaginary situations in which zombies run after the user, thereby encouraging her to run faster. The zombies are only heard as if they chase after the user. Similarly, traditional "audio walks" accompany the user by playing oral explanations on the scenes that are viewed by the user, taking into account her current location. The difference, however, is that "Zombies, Run!" augments any reality and is not bound to a specific route as the audio walks (see Schraffenberger and Van der Heide 2013).
3. Manovich warns us that such a link to our location may yield surveillance in scales unknown before.
4. For discussion on "free choice" in a technologically saturated environment, see Chapters 1, 7 and 8.
5. On the second section of this chapter, the major term used was "composite intentionality." In this section one can think of Verbeek's "hybrid intentionality" or "distributed intentionality" (Verbeek 2008b, 96) referring to technological intentionality that forms new associations with humans. The clip offers yet another

new form of distribution in which human intentionality is subjugated. We have seen this type of technological intentionality in the Cambridge Analytica affair, when a company used Facebook's profile system and the advertisement mechanisms to shape the users' political views and voting. The users experienced Facebook as shaping their wishes, desires, and fears. An analysis that remains faithful to human intentionality would regard the role of the technological platform as a means to an end and would seek to reveal the people operating the software. But the emergence of AI urges us to model situations that are not driven by human intentionalities behind the scene.

References

Ihde, Don. 1979. "Technology and Human Self-Conception." *The Southern Journal of Philosophy* 10 (1): 23–34.

Ihde, Don. 1990. *Technology and the Lifeworld: From Garden to Earth*. Bloomington and Indianapolis: Indiana University Press.

Ihde, Don. 2010. *Heidegger's Technologies: Postphenomenological Perspectives*. New York: Fordham University Press.

Latour, Bruno. 1992. "Where Are the Missing Masses? The Sociology of a Few Mundane Artifacts." In *Shaping Technology/Building Society: Studies in Sociotechnical Change*, edited by Wiebe E. Bijker and John Law, 225–258. Cambridge, London: MIT Press.

Liberati, Nicola. 2017. "Phenomenology, Poke´mon Go, and Other Augmented Reality Games." *Human Studies*. doi: 10.1007/s10746-017-9450-8.

Liberati, Nicola and Shoji Nagataki. 2015. "The AR Glasses' "Non-Neutrality": Their Knock-on Effects on the Subject and on the Giveness of the Object." *Ethics and Information Technology* 17 (2): 125–137.

Manovich, Lev. 2006. "The Poetics of Augmented Space." *Visual Communications* (Sage) 5 (2): 219–240.

Michelfelder, Diane. 2014. "Driving While Beagleated." *Techné: Research in Philosophy and Technology* 18 (1/2): 117–132.

Puech, Michel. 2016. *The Ethics of Ordinary Technology*. New York and London: Routledge.

Schraffenberger, Hanna and Edwin Van der Heide. 2013. "From Coexistence to Interaction: Influences Between the Virtual and the Real in Augmented Reality." Edited by K. Cleland, L. Fisher, and R. Harvey. *Proceedings of the 19th International Symposium of Electronic Arts*. Sydney: http://ses.library.usyd.edu.au/handle/2123/9475. 1-3.

Uricchio, William. 2011. "The Algorithmic Turn: Photosynth, Augmented Reality and the Changing Implications of the Image." *Visual Studies* 26 (1): 25–35.

Van Den Eede, Yoni. 2011. "In Between Us: On the Transparency and Opacity of Technological Mediation." *Foundations in Science* 16: 139–159.

Verbeek, Peter-Paul. 2005. *What Things Do: Philosophical Reflections on Technology, Agency and Design*. University Park, PA: The Pennsylvania State University Press.

Verbeek, Peter-Paul. 2008a. "Cyborg Intentionality: Rethinking the Phenomenology of Human–Technology Relations." *Phenomenology and Cognitive Science* 7: 387–395.

Verbeek, Peter-Paul. 2008b. "Morality in Design: Design Ethics and the Morality of Technological Artifacts." In *Philosophy and Design: From Engineering to Architecture*, edited by Pieter E. Vermaas, Peter Kroes, Andrew Light, and Steven A. Moore, 91–103. Dordrecht: Springer.

Wellner, Galit, 2013. "No Longer a Phone: The Cellphone as an Enabler of Augmented Reality." *Transfers* 3 (2): 70–88.

Wellner, Galit. 2015. *A Postphenomenological Inquiry of Cell Phones: Genealogies, Meanings, and Becoming*. Lanham: Lexington Books.

Wellner, Galit. 2017. "I-Media-World: The Algorithmic Shift from Hermeneutic Relations to Writing Relations." In *Postphenomenology and Media: Essays on Human–Media–World Relations*, edited by Yoni Van Den Eede, Stacey Irwin, and Galit Wellner, 207–228. Lanham: Lexington Books.

Wiltse, Heather. 2014. "Unpacking Digital Material Mediation." *Techné: Research in Philosophy and Technology* 18 (3): 154–182.

Zuboff, Shoshana. 2019. *The Age of Surveillance Capitalism: The Fight for the Future at the New Frontier of Power*. New York: Profile Books.

PART FOUR

REVEALING THINGS THAT REVEAL US

The idea of revealing things might at first seem rather unusual, perhaps conjuring up only a vague image of a newly designed product being triumphantly revealed at a product launch or industry keynote presentation. Yet this image points to the problems we now have with properly revealing things, in the sense of making present their basic character and operational capabilities. Things cannot be revealed just by pulling off whatever cover might be hiding them as they sit on a pedestal, waiting to be presented to the world. This is because they are, in a fundamental sense, not all there. Even though they can become embodied in physical form, what they are and do is constituted just as much, if not more, by connections they have to other resources, practices, and processes. Properly revealing these things requires some ingenuity. This is especially the case since they often are designed to conceal their operations from those who use them, or the things themselves conceal their operations from those who design them! Some things are designed to collect valuable data about those who use them that is generated during use and fed back into the underlying platforms and to others in the network; some things shape and bind themselves to humans in hybrid fashion, interweaving their operations within us in ways that are complex; in other things operations are submerged and seemingly inaccessible yet clearly present. They reveal us and our activities, sometimes in staggeringly precise detail or in oblique fashion. Yet we struggle to find appropriate methods for adequately and meaningfully revealing them and what they do in the context of our everyday lives and activities.

The chapters in this part all somehow engage with attempts to reveal aspects of things and their relations that tend to be hidden, inaccessible, or illegible to us as humans using them. They are exercises in revealing, but ones that are not naïve.

They all recognize the gaps between things and us, and the presence of withdrawn aspects that always partially escape us or cannot be translated into a form we can fully make sense of. At the same time, they make propositions for new kinds of conceptual and practical relations to things. They engage in various ways with the question: How can we reveal things that reveal us in order to relate to them with better understanding, intention, and agency?

10 IMAGINING THINGS: UNFOLDING THE "OF" IN PHILOSOPHY OF TECHNOLOGY, THROUGH OBJECT-ORIENTED ONTOLOGY

Yoni Van Den Eede

Introduction

As we try to understand what it means to "relate to things that relate to us," each of us automatically forms an image or concept in our mind of what those words mean: *relating, things, us*. We are working with terms and meanings that cannot help but be partially self-evident, because they are part of a long history—culturally and individually. Sets of meanings have been culturally sedimented and are reproduced in our individual upbringing. And yet, what we seek to do here is precisely to investigate whether those meanings, as concerns *relating, things,* and *us*, still hold, judged in the light of new technological developments that exactly overturn what we thought things, relating, and ourselves to be. But what if, even before we start transmogrifying the meanings of the terms, those terms have already implicitly steered us into a certain direction—by way of our pre-given understandings of them, but even more saliently, because they *are those* terms?

With "relating" and "things," we enter a long-standing debate in philosophy between the proponents of substance and of relation. For a long time substance thinking had reigned in the history of philosophy, until around the nineteenth and certainly throughout the twentieth century the relational view became dominant. Interestingly, commonsense thinking and discourse seem to run a bit behind the latest philosophical movement, rooted as they still are in the legacy of one of those last great substance thinkers, Descartes. We cannot seem to shake Cartesian dualism—still thinking on an everyday basis, if only implicitly, in terms of an "us" (a human subject) placed before a "them" (an object, either of nature or of our own making). All the while in academia, at least in the humanities and the social sciences, relationalism is all the rage, perhaps most conspicuously represented by Actor-Network Theory and by fields and approaches that go back to Deleuze and Guattari.

But is this a matter of common sense not yet having caught up with the "right" way of looking at the world? Or does common sense "choose" to cling to something in that repugnant Cartesianism that appears to be valuable after all? Maybe common sense is trying to tell us something. A tinge of this hunch clings to scholarship. Certainly, after decades of postmodernism, posthumanism, cyborg theory, and the like we have gotten used to constant disclaiming: "We know there are no such things (oops!) as subjects and objects. We are machinic, cyborgian, quasi-, assemblages and so on." And yet we have to use words. For that, we are left with linguistic tools that go back not decades, but centuries. But even more fundamentally, basic experience teaches that there *are* humans and things. We are us and they are them. I can hold a ball in my hand and clearly distinguish between me and the ball; I can toss it, damage it, and put it in a drawer. It is still me *and* the ball. All talk of assemblages notwithstanding, I am not that ball, and the ball is not me. Subjects and objects: at least they very much *appear* to exist.

In the contemporary philosophy of technology, this tension is particularly pertinent—though not often made explicit, as it extends into the field's very essence. Philosophy "of" technology necessarily wants to zoom in on *something*, namely, technology. As we will see, it does this paradoxically by letting its subject matter disappear in a way, by making this purported "thing" of technology scatter into innumerable constellations of relations. Indeed like most strands in academic research, philosophy of technology nowadays is dominantly relationalist. Then again, it still has to deal with "technology," say something meaningful about *it*. It wants to say something about it. Philosophy of technology pre-eminently struggles with substance and relation. A symptom of this conflict can be said to be its several "turns"—empirical turn, material turn, and so on—the movement between which can be seen as a continuous bouncing back and forth on the spectrum from relation to substance.

Notwithstanding the overall emphasis on relation in academe, in fact going against its grain, object-oriented ontology (OOO) has been working in the course of the last decade or so to reinstate substance to a pivotal position. Yet this entails no simple return to substance or plain reversion to commonsense dualism. Nonetheless, OOO does seem to pick up from common sense this hunch about "things being there." Philosophy of technology, despite its attention to concrete technologies, materiality, and such, has scarcely engaged with OOO—probably, we can speculate, because of the field's mainly relationalist orientation. However, what I will argue in what follows is that an OOO-inspired stance can shed light on (1) the general tension between commonsense substance thinking and the overarching relationalist paradigm of academic discourse and (2) its specific instantiation in philosophy of technology: its "paradoxical position." Once again, this should not imply a dumb return to older substance-dominated theories. Certainly also given recent technological developments toward more and more "interwovenness"—of humans and technologies, of cognition and artifacts, and so on—we should actually be *wary* of focusing on objects too much. Yet, exactly to acquire such a "not-too-focused-on-objects" stance, I propose philosophy of technology should look more—and only seemingly paradoxically—to OOO, especially as worked out by Graham Harman.

Object-oriented perspectives are often misunderstood as attempting to return to "objects," that is, in the Cartesian sense. Although OOO does set out to salvage substances, we should not interpret it *too* commonsensically. In fact, for OOO to "work," I will contend, we need to try and appropriate a *new kind* of common sense, learn to see the world "through the lens" of OOO, thus positioning ourselves in between currently reigning common sense and predominant academic relationalism, just as much as OOO finds a road in between undermining and overmining (see below). This new and still-to-be-deployed commonsense "paradigm" would circle around the idea of refusing to "know" and should compel us to an attitude of cultivated uncertainty. In itself, OOO does not offer something mysterious; it just argues that *things are* to a certain extent mysterious. Sometimes it is expected from OOO that it tell us something about objects that was heretofore hidden. Often in the context of, again, those developments toward "smart" technologies, algorithmic prediction, machine learning, and the like, OOO is brought in as a perspective that could help to make sense of concealed realities, created by technologies. But expecting the hidden to become unveiled by proxy of OOO is only *partly* reasonable. In a much more profound sense, OOO teaches us to *live with the hidden*.

First, I start by elaborating my point about the "betrayal of common sense." Then, I sketch the aforementioned issue in philosophy of technology: its thorny relationship with its "of." Subsequently, I introduce Harman's approach and link up his stance to the topic of technology, going into some existing approaches that put OOO in the service of thinking about technological matters. About the latter half of the chapter, then, is devoted to reflecting on the matter of "imagining things" problematized here, from the perspective of OOO. First I outline some consequences of OOO's premise of the inaccessibility of objects' substantive core, that are important for our discussion. Then I proceed to elaborate how this approach leads us in the end to the idea of the cultivation of a fundamental attitude of uncertainty. I illustrate this by a brief discussion of the novel *Klont* by the Dutch writer Maxim Februari. Finally, I conclude with some short reflections on rethinking common sense.

The Betrayal of Common Sense

Following Haraway (1991), we have learned to see ourselves as cyborgs; following Deleuze and Guattari (2004), as parts of assemblages; following Latour (1993) (who follows Serres), as quasi-objects; and so on. Scholarship in the humanities, especially throughout its many "turns" in recent years (digital, narrative, material, nonhuman, etc.), has done its best to acquaint us with an image of ourselves as certainly not freestanding, autonomous subjects, but interrelated with the things that surround us. That is meant of course in a fundamental, ontological way: the interwovenness constitutes us—and other things, for that matter.

But have we really *learned*? Everyday thinking and discourse tell us otherwise. We cannot help but reify, look at ourselves as apart from the world—*part* of the world, certainly, but reasonably independent from it at the same time. When it comes to technology, this shows itself in the instrumentalism that still characterizes much of our daily involvement with technologies, despite similar developments toward "becoming aware of interrelationality" in the philosophy of technology. It is a phenomenon that can hardly be delineated precisely, let alone quantified. But a quick look at our (own) everyday attitude toward technology generates plenty of examples: our apparent obliviousness concerning the problems of privacy on social media; the inability to see our individual usage of cars (and of fossil fuel in general) as connected with ecological crises; our willingness to prefer easy industrially processed foodstuffs over more healthy and environmentally friendly products (however more challenging to cook and consume); and so on. These may not always be instrumentalism pure and simple, thought through in its essence, that is, the conscious use of technologies as means to an end, with the expectation that they are nothing more than that; in other words, that they are "neutral." But an implicit line of thinking is present here that fragments the world into clear-cut elements—of which it is assumed they will act, work, move, or bend, with no resistance, according to our more or less explicit plans and purposes.

So, common sense betrays us. Stubbornly it stands in the way of a true embrace of the new relational worldview that scholarship hands us. Contemporary philosophy of technology indeed acts against this obstinacy of our at-face-value thinking about technology. It seeks to unmask the reified rendition of technology for what it is, namely, a reification of what is in actuality a much richer, much more complex situation of "networkedness." Instrumentalism is not able to account for that. But neither is its counterpart, determinism (or essentialism), which also entails a reification, but on the other side of the spectrum so to speak. Here technology becomes an all-determining, all-penetrating force—not a simple neutral thing or tool, but an all-encompassing thing, in which we are wrapped up without differentiation or capacity for meaningful action. There is a deep kinship in this regard between instrumentalism and determinism. A sign of this is the central role that the value of "efficiency" plays in both, or at least in important instances of both. For instrumentalism, technologies are just "efficient" means to an end. Their efficiency, or adequate technical functioning, is the only value by which we need to assess them. But (among others) in Jacques Ellul's (purported) determinism as well, efficiency represents the main value propagated by technology (Ellul 1964). However, here efficiency becomes an overarching steering principle, from which there is nigh no escape.

Contemporary philosophy of technology pierces through the cinema screen of ostensible, exclusive concern-with-efficiency. It demonstrates how beyond our limited understanding of technology in terms of efficiency, technology is constituted in fact by a broad array of networks, of relations, between all sorts of factors: social, economic, political, psychological, perceptual, cultural, hermeneutic, and so on. In essence, it oversteps a *narrow* notion of technology in order to arrive at a *wider* mode in which we grasp technology for what it really is: a constellation of relations (Van Den Eede 2016). For instance, Andrew Feenberg (2002) shows, in line with SCOT (Social Construction of Technology), how technologies come about as structures in which many values—social, economic, political—are at play and which, most prominently, instantiate and consolidate certain power relations and class struggles. Or postphenomenology, as worked out by Don Ihde (1990) and Peter-Paul Verbeek (2005) among others, analyzes how technologies are part of a framework in which humans and technologies "co-constitute" each other—the relation comes first—and how in this way perception, interpretation, and action are "mediated."

In this way, common sense in turn is betrayed. It goes toppling over, elegantly tripped up by philosophy of technology. We are inclined to equate technology with a *thing* or *things* or in any case some well-circumscribable entity or force. We are wont to do that; who does not instinctively think of objects such as computers, smartphones, machine infrastructures, and so on when the term "technology" is mentioned? But as soon as we do this, philosophy of technology barges in to mercilessly demonstrate that technology is not a thing. Or, more precisely, it is not *only* a thing or things; it is much more than a thing or things. Technology is not really technology, as we know it. It is all these other "things" as well.

Two betrayals. Yet, we should not think we are dealing here with just the plain matter of dualistic tension between common sense and philosophical-academic thinking—with the two vying for authority. This would be a too simplistic way of presenting the situation, which is much messier. At least we are using the "old" and the "new" language, concepts, modes of thinking next to each other. Or, we are intermingling them in confused and sometimes confusing ways. The old categories may have a stronger foothold in our mind since they are more deeply sedimented and entrenched. But the new categories—of interwovenness, (inter)relationality, intermediality, et cetera—have a vaguely familiar ring to them as they hint at experiences that we are starting to recognize. Illustrative of this are headlines as well as the general tone of articles in the popular press that ask, for instance, whether, due to technological developments, we are becoming "a new human being," or whether our children "will be cyborgs."[1] The question is: do we frame these issues already *in a certain way* because of the terms we use, namely, of "us" becoming "this" or "that"? On the basis of which idea of the human being do we first of all begin to wonder about these developments? No doubt it is still for a large part our modern, Cartesian intellectual legacy that is raising its voice, despite, once more, so much work done in philosophy and other disciplines to subvert this legacy.

However, importantly, this also affects scholarship itself. We can expect that "us scholars" are still working with(in) the old frames just as much—perhaps with a little bit more awareness, and maybe in somewhat more hybrid forms, but nonetheless, these frames are there, at least to some extent. To put it provocatively: to which extent does this vestigial subject-object thinking cloud or distort our capacity for really piercing through to the purported reality of interrelationality? To be even more provocative: seen from this angle, are we, "within" scholarship, actually (already) equipped enough to understand and work with our own newly fashioned insights about interwovenness?

The "of" in Philosophy of Technology

We are faced with an intriguing kind of perplexity. And a variant of this haunts philosophy of technology. Somehow it is even hiding in plain sight: namely, in its "of." It is as if the aforementioned struggle and mix-up between common sense and academic relationalism must be parsed through its infrastructure in some way. On the one hand, the field, as we saw, has amply shown how technology is much more than a (mere) "thing." On the other hand, however, if only because it is or calls itself philosophy *of* technology, it retains a link to commonsense thinking about "things."

Philosophy of technology in its contemporary incarnation has in the last decades become something of an institutionalized philosophical (sub)discipline, with its own organizations, journals, conferences, and even dedicated university departments and research groups—like before philosophies of history, of science, of biology, et cetera came into being. One could pause at this relatively recent (nineteenth- and twentieth-century) evolution of there emerging philosophies *of*… along typically late-modern lines of specialization. Apart from that, there is a certain curse to the preposition "of" in this context. To create and to have a philosophy *of* something obviously enables one to zoom in on an issue that heretofore was perhaps treated in a stepmotherly way or not conceptualized at all. But it also entails that whatever is done and said within the field is somehow strictly to be done and to be said *within the field*, that is, within the bounds of the "of," that double-edged sword of conceptual reach *and* constraint. No matter how far one drifts off, in developing the philosophy *of…*, from the subject matter at stake—the short description of which is meant to follow the "of" (technology, science, biology, and so on)—one always eventually has to return in some way to "home base"; be a good girl or boy and make some useful statements about the subject matter. Because we *are* "doing," after all, philosophy *of*….

An unspoken suspicion of this uncertainty lingers through some of the central debates in the field. Contemporary philosophy of technology emerged in a reaction to the so-called "classic" philosophy of technology, represented by figures such as Ellul and Heidegger. These saw technology, it is said, as an overarching force or principle that affects and penetrates eventually all domains of life. In opposition to this, philosophers of the "empirical turn" (Achterhuis 2001), from the 1980s and 1990s onward, wanted to look at concrete technologies, in specific use contexts—to find that technology is not the all-determining force classic philosophy of technology took it to be, but can be modified, adapted, reappropriated, contested, and so on. The empirical turn thus reacted against the determinism of the older "wave"—which had been in itself largely a response to instrumentalism: the instrumentalism of scientism and positivism, and the already in those days (first half and middle of the twentieth century) popular belief that technologies could unilinearly solve problems. Not that empirical turn philosophers proposed a simple return to instrumentalism; they sought to find a middle road between the two, arguing that technologies take on a bit of both. Technologies do have effects; that is, they are not neutral. However, they are also not monolithic structures that cannot be changed; that is, we can at least to some extent control them.

The move past "efficiency," which I described in the previous section, has been central to this endeavor. At the same time, paradoxicality clings to it. In order to delineate its object of study—technology—contemporary philosophy of technology has had to lose it. The so-called thing of technology disperses into a web of relations. Right as it appears, it vanishes. This is not often problematized. Nevertheless, an implicit realization of this conundrum has perhaps been driving some of the (other) movements and discussions in the field. The "material turn," a phenomenon in the humanities as such, has also pervaded philosophy of technology. (We will see Graham Harman point out how the movement of the so-called "new materialism" is in fact not at all about *matter*, but rather about relations, social and historical construction, and so on.) In philosophy of technology, the material turn, which is closely linked to the empirical turn, meant to put the focus on material practices and things, instead of on human understanding and discourse, thus reacting against—as all turns react against something—the humanities' decade-long occupation with linguistic structures and phenomena. Yet lately several authors have been "re-turning" from this turn, trying to correct for its own one-sidedness and risk of overemphasis. Mark Coeckelbergh (2017b, 2017a) wants to renew our attention to language and discourse. Verbeek suggests we need "one more turn after… [the] material turn" (2015, 192), refocusing again on human meaning-making processes.

All this searching and changing of directions—the very concept of a turn—can be said to characterize many scholarly domains and disciplines. Yet this overall ebb and flow of *now looking here, then looking there*—and then repeat, and so on—may also testify to a deeper-lying issue. Dominic Smith (2018) rightly questions the dynamic of "turning" as such. For him in the end the basic problem is the empirical turn that sets us up with certain methodological and conceptual restrictions, and he calls on philosophy of technology to engage anew with the transcendental—which the empirical turn sought to block out and discard, wanting to account only for empirically observable phenomena. Smith's analysis forms part of a bigger debate on the status of the empirical turn (Scharff 2012; Smith 2015b, 2015a; Lemmens

2017; Van Den Eede, Goeminne, and Van den Bossche 2017), in which several other authors have suggested to revive at least some notion of "the transcendental" in our thinking about technology. According to some of these commentators, the empirical turn has been pushed too far, once again too much in the direction of a newfangled instrumentalism. That could leave us unable to account for wider, more encompassing phenomena that are not empirically mappable or graspable—such as comprehensive political or ontological conditions.

Clearly, we wrestle. In what follows I want to work out my own approach to the issue, specifically starting from OOO and reading the "of" conundrum in the philosophy of technology as an instance of the substance-relation conflict. The shifting and browsing across turns is, in my reading, a symptom of this. (By way of which I do not want to question the turns and corresponding proposals in themselves as meaningless. They make for authentic and valuable projects that are probably needed in the development of a field. But beyond that, they may also point to dynamics that play out on other, higher levels of abstraction.) We go to sufficient lengths and are at pains to let technology disseminate into myriad relations and networks. Technology is certainly not "technology." And subsequently, the bungee cord bounces back, and we feel obligated once more to pay our respects to the "of"—as if to assure ourselves and others: "But we are still looking at *technology*, no worries." Is there a way to reconnect the two sides?

OOO and Technology

It is not my intention to offer an extensive introduction to object-oriented ontology and related fields here (good introductions are widely available online and in print[2]), but in order to frame my argument, I need to briefly sketch its contours. I zoom in mainly on the work of Graham Harman, who can be seen as OOO's helmsman. Crucially, "object-oriented" in Harman's sense is not what we often think it to be. The term evokes misunderstanding, as if it involves a throwback to modern categories. Object-oriented ontology is paradoxically not about *objects*—not as we *know* them, at least. *Feeling* Harman's "object standpoint" brings us somewhere else altogether.

Harman starts out reading Heidegger in a completely innovative way. The tool analysis from *Being and Time*, which distinguishes between presence-at-hand and readiness-to-hand, according to Harman (2002), is not about literal tool use, or at least, not only about that. In the common interpretation, Heidegger denounces the objectifying, conceptualizing attitude of Western metaphysics (presence-at-hand: picturing the hammer as hammer object, placing it before us), pointing out how primordially, before we start to objectify, we are always already engaged in a relational network (readiness-to-hand: we use the hammer in work, in which it "disappears"). (Notice the parallels with contemporary philosophy of technology's central "moving-beyond-efficiency" endeavor—to which I will return.) Harman does not contest the dichotomy as such, but turns its standard reading upside down. Readiness-to-hand equates with *substance*, not with relation. This substance escapes all relation—any direct perceptual, practical, or theoretical grasp whatsoever. Initially Harman called this inaccessible substantive "core" *tool-being*, but he seems to have largely left this (indeed slightly confusing) denomination behind in favor of the term *real object*.

Subsequently Harman brings in another dichotomy, from Husserl: the distinction between a sensual (or intentional) object and its plurality of traits or qualities. Taken together, these two dichotomies then go on to make up Harman's in the meantime familiar "quadruple object," composed of four quadrants: a real object, real qualities, a sensual object, and sensual qualities. Between these, all sorts of links and relations are in play (Harman 2011), but a fundamental asymmetry reigns between the real and the sensual. As such, "[r]eal objects can touch only through the medium of an intentional [or sensual] object, and intentional objects can touch only through the medium of a real one" (Harman 2009, 208). Real objects as well as real qualities can never be grasped directly.

In recent years Harman has been further elaborating his theory, often in conversation with Actor-Network Theory (ANT), especially Latour's work. Harman sees Latour as one of the most important contemporary thinkers (he has dedicated no less than two full books to his work: Harman 2009, 2014): someone who, like him, goes beyond anthropocentrism to devote attention to the countless objects making up our world. But in Harman's view Latour is guilty of what he calls overmining, which means reducing objects "upward" to their effects, relations, historical contexts, and the like. Undermining, on the contrary, is reducing objects "downward" to their constituent parts or elements, as for instance atomism does. For Harman, there is an in-between (the "third table," as described in the essay of the same title: Harman 2012). This is the real object: neither constituent element nor relation or effect. Latour refuses to account for this hidden substantive core, which in Harman's view is a very much needed component for the theory—not just a whim: "Unless the thing holds something in reserve behind its current relations, nothing would ever change" (2009, 187). The real object is what stirs things up, so to speak; it is what makes for change.

As Harman argues (together with Levi R. Bryant), the "new materialism" movement of which ANT can be said to be a part does not have much to do with the "tiny material particles" (Harman 2018, 135) of older materialisms. This new kind of materialism that focuses on material practices and how these come about in relational networks—in the process, thus, reducing things to relations—is an overmining materialism (Harman 2016, 15). "Actor-networks are simply the inverted form of atom-networks" (ibid., 19). For his own alternative approach, Harman proposes the term *immaterialism*. This is akin to ANT, but applies a couple of different emphases, all flowing from that one important decision to retain a real object that escapes all relation. When it comes to looking at objects, how objects relate and, crucially, how they come together to form new objects, OOO will according to Harman not so much stress controversies and change as ANT does, but rather also look at "the moments of uncontroversial reality in things," at "simple success and failure" (ibid., 40) and at instances of what Harman calls, borrowing from Lynn Margulis, *symbiosis*. I will return to this in more detail throughout the next section.

Before we go on, we need to ask: Where is technology left, if the tool analysis is not in the first instance about *tools*? Indeed OOO "flattens" ontology, claiming that all things in the world are on exactly the same ontological footing. Technology in that regard is nothing special. It is just another "thing"—though of course not in the classic Cartesian sense (as we will also still see in more detail). This does not preclude that we can distinguish between different sorts or categories of objects.[3] *Ontologically,* however, all have the same status. Regularly, Harman refers to a technology or technologies in illustrations, but he has not (as of yet) specifically investigated technology within the bounds of his framework.

This might be one reason why Harman and OOO have up until now been largely neglected by contemporary philosophy of technology, a few exceptions notwithstanding. One systematic attempt to put OOO to work for the study of technology comes from Matt Hayler (2015), although his project is more situated within cognitive science. Hayler defines technology use as an encounter, and in his view, technology really becomes technology when it is used with a certain amount of skill. OOO for Hayler can help us to understand this process in which technologies have surprises in store, and a user who hones her skill gradually gets acquainted with more and more heretofore hidden aspects of the thing. This is immediately where Hayler deliberately deviates from Harman. Harman denies the possibility of an "asymptotic" account of knowledge, according to which we would be able to get progressively closer to reality. The real object stays hidden, full stop. Hayler, by contrast, consciously allows for a definition of knowledge as "an asymptotic edging-towards that is reflected by the success and repeatability of an activity" (ibid., 200). There are some problems with this (see below), but in any case, Hayler executes the useful exercise of importing the notion of substance into our thinking about technology, thereby supplementing approaches like postphenomenology that, as he points out, stay secluded to an involvement with relation alone. A related but different project, however, based on a similar sensibility, can be found with Thomas Sutherland (2013), who seeks to question what he calls the dominant "metaphysics of flux" typifying social theory, which epitomizes networks, relations, flows, and so on. This view unjustly favors becoming over being, he argues, but most crucially it has nontrivial consequences for thinking about political agency, as it tends to confirm the status quo rather than enable political praxis. Sutherland uses OOO to correct for this lack, to help boost "speculative reason."

Still, the amount of attention from philosophers of technology for OOO stays relatively modest. Why have the two not drawn nearer to each other? We can hypothesize about this. The involvement with "objects," the common theoretical sources in phenomenology (at least as far as postphenomenology is concerned), and the interest in the "hidden" (philosophers of technology widely share a sensitivity toward the accounting for implicit side effects, and so on)—these should warrant that the two fields sniff each other out much more. Yet perhaps there are simply too many compatibility problems. Harman's deliberately provocative defense of substance—a highly contentious notion of course since the "relational turn" in intellectual history of the twentieth century—might not gel after all, as Hayler's account helps to suggest, with the field's mainly relationalist bend. Or, OOO's fundamental-philosophical orientation, openly "metaphysical" in spirit, may be looked upon rather unfriendly from the perspective of empirical turn approaches. Of course Harman does use examples and at times even works out elaborate case studies. The fictional essays in *Circus Philosophicus* (2010) and the extensive study of the Dutch East India Company in *Immaterialism* are cases in point. But these often-playful explorations might be too far removed from philosophy of technology's preferred engagement with matters of technology in particular. As such, the "flat ontology" proffered by OOO, that puts all things—including humans and immaterial things such as ideas—on the same ontological level, may not sit comfortably with philosophy of technology's central project of trying to delineate what *technology* precisely *is*. But in fact, that tension is one of the main issues I want to tackle here. So, let's see if we can push this up a notch.

Consequences of Inaccessibility

I want to first outline a couple of traits of Harman's OOO that are specifically pertinent to our discussion here. Once more, we need to emphasize that the object-oriented stance is not a return to modern, Cartesian categories. But clearly, we have seen, neither is it a corroboration of the newer relationalisms. It offers an in-between—although also not in the sense of a combination, in that it would combine undermining *with* overmining, which Harman calls duomining. No, the in-between is a kind of no man's land, or better, a forever being-underway-toward. Real objects cannot be known, grasped, perceived—they withdraw; they escape, everywhere and always. But, we *have* a desire for knowing them. So we are always underway toward them. This is the ground attitude of philosophy, Harman points out. Now, the premise that all objects partly escape us can seem like just a trivial proposition. However, when following through its consequences for how we think about things—things meant here in the generic sense of how we look at and deal with the world—we arrive at a couple of interesting observations. Specifically for philosophy of technology, I believe certainly the following characteristics of the theory are at stake[4]:

1) Uncertainty
   We never really know objects in their core. For Harman this goes back to the Socratic attitude. Socrates neither undermines nor overmines. "[W]hat Socrates seeks is not a kind of knowledge, since he is interested neither in

what virtue, justice and friendship *are made of* [undermining] nor in what they *do* [overmining], though this has often been forgotten" (Harman 2018, 47). Socrates claims to know nothing, and this is not empty posture. He is always underway toward knowledge. Harman reminds the reader that "the original meaning of the Greek word *philosophia* is not knowledge and not wisdom, but the *love* of a wisdom that can never fully be attained" (ibid.). Philosophy, with an expression by Nicholas of Cusa, "is nothing if not the permanent practice of 'learned ignorance'" (ibid., 176). Rather than clinging to the notions of knowledge and truth, we should concern ourselves with *reality*—of which there is no direct knowledge ("*there is no direct knowledge of anything*": ibid.; original emphasis), that is, a definition in literal terms. We can approach reality indirectly, through metaphor for instance (later on Harman does outline a concept of knowledge, however one that, again, cannot entail a direct grasp of the real: see ibid., 185ff.). This recognition, or even better, embrace of uncertainty is central to the object-oriented stance.

2) Objects Come Together to Form New Objects
   For OOO aficionados this is obvious, of course, but it deserves stressing just to point out the "unusualness" of the perspective compared to either Actor-Network-like overmining or—for one—atomistic undermining. We should not imagine "objects" in the object-oriented sense as pre-existing entities that then form a relation with each other, end of story. Harman posits quite strongly that "any relation between separate things produces a new composite object" (ibid., 167). Objects continuously come together to form composite or "compound objects." This goes against the grain of ANT: "ANT would reject [...] the notion that compound entities are new things-in-themselves rather than just transient relational events" (ibid., 107). Objects for OOO come in and go out of existence all the time, although some are more visible than others. With his analyses of things such as the Dutch East India Company in *Immaterialism* and of the American Civil War in *Object-Oriented Ontology*, Harman means to illustrate how, indeed, these are *objects*: compound objects composed of countless other objects, but still objects in their own right. Whereas a relationalist approach such as ANT would certainly regard these events solely as events, "for OOO every real event is also a real object" (ibid., 53). Remember that objects can be material as well as immaterial, solid, ephemeral, … There is only one requirement for something to be an object, and that is "that it be irreducible in both directions: an object is *more than its pieces* and *less than its effects*" (ibid.; original emphasis). This relates to the first point about uncertainty. The status of uncertainty connects essentially to what objects are, and vice versa.

3) Creation of New Objects instead of Discovery of the Withdrawn
   The creation of new objects happens all the time—much more perhaps than we would implicitly expect, given that we are accustomed to

thinking about objects in the way we do: as relatively fixed, self-sufficient, independent entities. This takes some getting used to. Especially this has consequences for thinking about the new technological developments around algorithms, machine learning, and so on. In the debate and literature on these, often "the hidden" is a central theme. These kinds of technology, it is said, create conditions in which something goes lurking behind the screen of appearance. We see our social media newsfeed, but the algorithms that create it—that determine what we see and don't see, and why—stay hidden "behind the scenes." It would seem that an approach like OOO is perfectly suited to making sense of such a situation, in which a constellation of (digital and material) objects is shown to us in a specific way, from a certain vantage point, while something else escapes. But here is the catch: for OOO, whenever something is "revealed," that means, whenever a new relation is forged with something, it stops being the object it is; it turns into a new object. What is more, revealing of the ontologically withdrawn—the real object, the substantive core—is impossible. Turning this around: when something is "revealable," then that must mean it is not really the *real*—for the real stays concealed, full stop. One can have an indirect relation with a real object, of course, for instance, through metaphor (see ibid., 61ff.). But then, again, a new object arises (see ibid., 88–89). What should be the relevance of all this for thinking about the "hidden" in digitally mediated/constructed environments? It is a kind of slap in the face for us who are groomed to think about revealing and concealing in a certain way. The endeavor of uncovering the hidden … is a skewed way of putting it; rather what we should be concerned with is the *project of making new objects*. Instead of asking Facebook to lift their cloak, to tell us their secrets, we could say: let us make a new object "Facebook," that has a different revealed-concealed structure. It may be a banal difference, this manner of speaking. But it puts us in touch, lets us become acquainted more with the fundamental character of uncertainty bequeathed to things.

4) Objects Have Life Trajectories
Objects come into being by forming new, compound objects with other objects. From this it must logically follow that objects have a "life history." Harman has attempted from *Immaterialism* onwards to provide an account of how objects are "born," how they grow, flourish, decay, and finally die. Once again in reaction to ANT, he develops the notion of symbiosis, adapted loosely from Lynn Margulis's endosymbiosis. Symbiosis is the forming of a bond with another object, but not *just* a relation: a meaningful relation that defines the object as what it—from that moment on—is. This is what the notion of symbiosis delivers beyond ANT. For ANT, all relations have the same stature, and they are reciprocal. Symbiosis, however, refers to "a special type of relation that changes the reality of one of its *relata*, rather than merely resulting in discernible mutual impact" (Harman 2016, 49). It is often non-reciprocal. The event

of reading a life-changing book by a certain author, for example, will impact heavily upon the person reading the book—it creates a new object "person changed by the reading of the book"—but mostly not upon the author (although there may be instances in which a bond gets formed between a reader and author that affects the author as well). Also in opposition to ANT, Harman argues that symbiosis is not necessarily about visible success. "[A]ny theory that overmines objects by paraphrasing them in terms of 'what they do' [as pre-eminently ANT does] has already conceded that history is a roster of winners, devoid of undeserved success and undeserved failure" (ibid., 99). Harman, by contrast, also wants to devote attention to "dormant objects" (ibid., 42), to "the dogs that did not bark, or the barking dogs at moments when they slept" (ibid., 40). Symbiosis may come about in relative seclusion, may not be immediately visible for the object's environment. But still something is happening; the moment is part of the life trajectory of the object. (Notice the interesting connection with Ihde's notion of technologies having "trajectories" and "shelf lives": see Ihde 2017.[5])

In all this, we should not forget that these dynamics play out among all objects, among each other—possibly without any human intervention. For our purposes here, however, it does not harm to slightly favor the human standpoint. We are concerned in the first instance with *how to go about* thinking technology—and from there on to explore how to relate to things that relate to us—in the framework of philosophy of technology. This might seem a bit counterintuitive as well. Are we not adopting the object-oriented perspective precisely because we want to make sense of "smart"/learning/algorithmic things, that seem to be acquiring a sort of "life of their own"? Yes, however, in the context of this discussion, our first priority is the implicit image of "things" we are using in that exercise. How to "imagine" things? In the following section, I outline my proposal for an approach.

Cultivating an Attitude of Uncertainty

Let us reiterate: philosophy of technology thinks the technological "thing" as a set of networks and relations, and hence "loses" the "thing," about which it nevertheless wants to say something. This is a mirror image, an instance, of a more general friction—and regular confusion—between our commonsense thinking about things and widespread academic relationalism. We want to say something about things and at the same time we feel we cannot, because there are really only relations, right? Could we unfold philosophy of technology's "of" so that its "paradoxical position" becomes less of an issue?

Harman's OOO acquaints us with a totally new way of thinking and thinking about objects, if we care to embrace and try to "feel" it, see the world through its lens. There are objects. Yet, we cannot be *too sure* about them. Certainly we need to be careful about imagining "objects" as we are used to—lest we be unjustly *imagining* things that are not there. It's about the right way of imagining. That kind of imagining is (positively) tainted by incertitude. We are always underway toward knowledge of the real, but never reaching it—because it cannot be reached. Also, objects are not a static, once-and-for-all matter. They come together, form compound objects. New objects come about in the blink of an eye. And, objects have histories; they change and evolve, possibly by "taking on board" or "shedding" other objects, or by forming meaningful (or less meaningful) bonds with other objects. The world is a pandemonium of such contacts, clashes, melds, mergers. And each time the real object stays absolutely inaccessible and withdrawn—a mysterious pool of "extra," from which the new may arise.

Most crucially, OOO incorporates relation *and* substance. Could it thus provide us with a way of not having to choose between "things" and relations? I believe it can. What is more, given OOO's origin in Heidegger's tool analysis, it can shed new light on philosophy of technology's core dichotomy between a (narrower) efficiency-oriented and a (wider) network-oriented mode. The OOO approach helps us warm up to the idea that, perhaps, we have been looking at the distinction between the two modes too sternly.[6] Philosophy of technology puts aside "efficiency" in order to get at the "actual" situation of ubiquitous "networkedness"—but it does so too quickly. It seeks to move fast beyond the simplistic instrumentalist thing-as-means understanding of technology—wanting to betray common sense—with the aim of showing how "technology" in fact disperses (and disappears) into multiple relational networks. At the same time, philosophy *of* technology still has to account for a "something" that is technology. Common sense comes creeping in again through the backdoor—betraying us in turn.

OOO tips us off to the possibility, or even more strongly, to the indispensableness of combining the two "sides," just as Harman combines substance and relation. (Too superficial interpretations of his work might suggest it is all about substance, but this is really not the case—exactly the nifty combination of both constitutes its unique selling proposition, to put it somewhat disrespectfully.) Common sense, still reasoning in terms of objects that can be clearly delineated, has a point. But so does the relationalist paradigm. We should not be surprised if we appear to perceive and interact with things. Things are there. We just never access them directly, in their core. We ourselves are an object just as much, engaging into relations with other objects. And objects interact with each other in the same way. Due to the dazzling multiplicity of objects—not only relations are ubiquitous; objects *and* relations are—this leads to what could be called a *perspectival approach.*

The term "object-oriented" may seduce us into implicitly picturing a stance that is aimed at objects, but the term is misleading in this regard. More than a question of being aimed *toward* something—which would suggest more of an external viewpoint—the object-oriented approach is about "looking" *out from to* .... This does not lead to a purely internal viewpoint, but rather a mixture between internal and external: it is a from-inside-to-outside stance. It is taking the perspective of objects—and realizing that "our" (i.e., human) perspective has the same structural characteristics. Those structural characteristics are, simply, the fourfold object, made up of the two dichotomies (object-qualities; real-sensual). This entails that both dichotomies cling to all "perspectives." Especially the distinction between presence-at-hand and readiness-to-hand is relevant to our discussion here. The common praxical interpretation of the tool analysis, Harman argues, suggests the tool analysis to be in the first instance about tool use. In practical involvement, the tool disappears from view, is ready-to-hand. When it breaks down (gets broken or lost) we are suddenly forced to objectify it, conceptualize it as "standing before us": it becomes present-at-hand. The ready-to-hand situation is primordial, and relational. In the present-at-hand mode, by contrast, we reduce the tool unwarrantedly to a substance, in the spirit of (misled) Western metaphysics. The move of philosophy of technology from simplistic efficiency-orientation to wider network-orientation is in essence a Heideggerian one—back from everyday reification to "how it actually is": things are enveloped in the "relational totality" of "equipment."

Harman turns the tool analysis upside down and widens its scope. For him readiness-to-hand is about substance, while presence-at-hand is about relation. Only through breakdown, that is, relation, can something become present. "[I]ndividual objects are smothered and enslaved [namely, by the substantive core, into which they withdraw], emerging into the sun only in the moment of their breakdown" (Harman 2002, 45). At any time, objects are submerged into tool-being (i.e., readiness-to-hand), only "making contact" with other objects through breakdown, thus relation (i.e., presence-at-hand). That means that all object-object interactions have the structure of presence-at-hand and readiness-to-hand intertwined. "Heidegger tells us no more and no less than this: 'all reality has the structure of the tool and its breakdown'" (ibid., 67). The perspectival image helps to make sense of this: we are always looking *out from... toward* something (the "we" in this phrase is a generic "we" that may encompass any object). Heidegger's hammer example of the tool analysis points out how, in hammering, we do not notice the hammer as hammer "object" (i.e., objectified); we are instead focused on the work to be done, building a cabinet for instance. What is often too quickly disregarded, in referring to the hammer case, is that in the work we are still *focused on something*. Of course, the hammer is the subject matter in the analysis. Obviously, we observe (or actually in practice experience rather unconsciously) how the hammer disappears into equipment, instead of being objectified. And so we might be inclined to leave it at that, concluding: it is all really a matter of relational constellation. However, notwithstanding the hammer and surrounding equipment disappearing into relationality, *something else* is still being made present. We visualize the cabinet we are constructing, or we focus on the job of getting that painting up. Presence *and* equipment are there, entwined.

Following Harman's upside-down-turning and widening of the tool analysis, we then have to try to imagine not just a situation in which tools are used, or something practical is being done—philosophy of technology indeed often tends to limit itself to cases like that—but how the world is composed of such substance-relation interactions. The world is made up of myriad "perspectives" looking-out-from-toward. That also means: the efficiency-network tension adheres to all perspectives. It is not just a matter of shoving aside efficiency-orientedness and moving to network-orientedness. We can stop "turning" and shifting from one to the other "side" (and back). The two are always combined. Only, we have to account for and make sense of *how* they are, exactly.

The elaboration of "consequences of inaccessibility" above gives us cues. Most importantly, there is uncertainty. The inaccessibility of the real is a given, if we want to follow OOO. This makes Hayler's approach of technology use as ever-growing expertise at least partly problematic. Hayler, against Harman, allows for the possibility of an asymptotic movement toward the object: growing knowledge. But to allow for this is a logical inconsistency within the OOO framework. Only indirect knowledge is possible; uncertainty can never be expelled, and hence neither can we shake off the call and duty to do philosophy, as the eternal striving toward wisdom. The asymptotic view, by contrast, would suggest there is something like a "better," that is, truer knowledge possible, closer to the real than other, less true knowledge. Hayler's account of expert use is worthwhile in itself, but from OOO, one must at least also draw an altogether different conclusion when it comes to "approaching technology": abiding to the verdict of uncertainty, we actually have to reckon with the condition of things fundamentally *not nearing* to each other. Sutherland gets this right: "Object-oriented ontology is based upon the fundamental premise of alienation—objects, including humans, are inherently and ineluctably alienated not only from other objects, but also from themselves" (2013, 18). Indeed the "fundamental premise of alienation" is a *premise*: one accepts it or not, but if we accept it, there are implications.

The question is: How do we *live* this,[7] certainly with regard to "technology," our first concern here? There is only one option, and that is *cultivating* this attitude—which may be a painstaking undertaking. Telling ourselves constantly to "not know," *not* imagine that we can "imagine" "things," seems like an excruciating task. However, art and literature—as always having a bit more poetical leeway—can help.

As one illustration, the latest novel by Dutch writer Maxim Februari, *Klont* (2017), is useful in this regard. *Klont* is about life in a highly datafied world. The term *klont*, which translates as clump or lump, denominates the mass of data and algorithms that is taking on a life of its own, beyond human control. Or at least, that is one assumption. There is a lingering, hard-to-pinpoint unclarity about the clump. Throughout the whole story, the characters are trying to find out what it actually is. Rumors abound. The clump is never really "there," always seemingly out of conceptual and perceptual reach. Some people are unsure about whether they should even care about it. Others are truly concerned. Gradually, the contours become more and more visible—be it still rather elusively. On being asked what the clump is about, the editor-in-chief of an established magazine answers reflectingly:

> "It is a metaphor. For a world in which we have lost our choices." Or rather not a metaphor, but a truly existing interplay of powers. A worldview that offered more room to notifications about existence than to existence itself and that ruled compellingly. (ibid., 198; my translation)

One of the protagonists tries to make a living off claiming to know what is happening. Alexei Krups, a TED-like public speaker, is acquiring world fame with eloquent though slightly ominous talks about the clump. Despite the self-assuredness that befits a man in his position, he is trying to figure it all out as he goes along; just as everyone is doing, really. However, he might be the one—at least as much is suggested by the narrative—who is most approaching "how it really is." Krups in the story is a kind of OOO-inspired philosopher of technology, I might provocatively suggest. At one time he gets an epiphany, a flash of realization: an insight of the order "we have been looking at it the wrong way." And he observes, having come to a conclusion, "Technology did not consist of electricity and things tied to each other with wires, but of images of the human being and views on existence" (ibid., 149; my translation). But the new way that is needed to actually understand what the clump is, is, again, not yet within reach. Or in any case it cannot yet be introduced or communicated on a broad scale, Krump observes.

Read otherwise, in the terms of my argument here: the new notions and frames, needed to get a grip on the phenomenon, are not yet part of common sense, and hence some disconnect must remain. But rephrased positively, that entails: we have to embrace uncertainty. Indeed, throughout the story there is always the implicit suggestion, the feeling that our current frameworks do not suffice to make sense of the clump. What is more, our frameworks might even close off, beforehand, an accurate understanding. At the risk of perhaps spoiling the plot too much (the reader be warned at this point), in the end it appears that "life goes on"—just like that. The characters go on, in multiple senses of the word (practical, existential), having apparently missed the opportunity to develop a truly adequate notion of the clump. The clump is still there, but life as it happens, with all its psychological, social, political wrangling and tug-of-war, "wins out," at least as far as day-to-day worries and activities are concerned. Extending this, one could say, common sense "wins"—after all? Yet, right until the end, that persistent, stingy suggestion remains that we are missing something, exactly because of our assumptions. The hard work of changing the grounds on which we think, and experiment with other ways of looking, but *really looking*—like looking *from-out-of*—is still there, lying in wait. This is essentially a reminder as to the uncertainty we must cultivate with regard to things as such. And this may be the less obvious underlying message of the narrative. We *ought* to be unsure: unsure about things as such.

This may seem a gloomy assessment overall. The assertion that *we can never really know* may feel like a heavy cloak of intellectual and even existential deficiency landing down on us. Yet there is definitely an upside. We can be *safely sure* about our unsureness. We can revel in this paradoxicality—like we can marvel at the paradoxicality of technology being a thing and a constellation of networks at the same time.[8] Yes, we may endeavor to get better and better at things, at using technology, loosely à la Hayler—there is nothing wrong with that; however, we should not expect a gradual coming closer to the truth. Rather, we should primordially nourish the eternal remainder of alienatedness, the impossibility of ever becoming 100 percent sure. For there is a silver lining around this "cloud of unknowing." It comes from the mind-blowing *multiplicity* of things. We lose absolute knowledge and absolute certainty, but we gain the rise of ever-new and ever-more (possibly compound) objects, with their own and interlocking life histories, their often-dormant aspects, but also possible surprises. Instead of truer knowledge being achieved, a *new object* comes about. We can find pleasure in this constant creation—like a god overseeing the rumble of things coming in and going out of existence, and everything in between, however, without the power of omnipotence. We are not always the maker, though surely in the case of technology we often are—yet the viewpoint developed here should at all times remind us of there being a "world out there" (the *real world is out there*, instead of the truth, as in *The X-Files*).[9] Imagining things entails also trying to *imagine* this world, taking the perspective of (other) things, not forgetting that the substance-relation dynamic, or, transplanted to philosophy of technology, the efficiency-network dynamic, characterizes all particular from-out-of structures.

Conclusion: A New Common Sense?

We began with *asking about the project of asking about* relating to things that relate to us. The technological developments under scrutiny—of digitization, "smart" things, nanotech, et cetera—already compel us to reconsider what we think a thing is, and how "we," that is, humans, position and define ourselves in relation to it. But what if, seen from a certain perspective, our framing of the issue in these terms exactly hinders a clear view on it? Digging deeper we must ask, *before* those questions become possible to pose—"what is a thing?" "what are *we*?"—why we actually use these terms. And what they do with our way of thinking. And to what extent they might be so ingrained in our thinking that we hardly notice the immense difficulty we have in shaking their implicit influence.

Yet, how could we ever *not* start from our presuppositions, either commonsensical or academic? Could we first change our idea of what it is to be a human being, and then (re)consider the "developments"? Correspondingly, could we learn to see things not as things right away, but as other "stuff" in the first place?

What is left for us to do, then—playfully, if you will—is to find ways, or at least probe possibilities to *rethink commonsensically* our bond with technology. Let academic insights in that process inspire us as much as they can. In a way, all philosophical scholarship always aims to question, even subvert, common sense. The same goes for Harman, and the same goes for my adaptation of OOO in the service of philosophy of technology. But obviously, as long as the new insights do not trickle through to common sense, common sense "as is" remains firmly fixed in place—laughing us in the face as it were. Some people would reason from common sense that the aforementioned brainchildren of Graham Harman are merely frivolous postures. How could ideas so far removed from commonsense reasoning ever begin to transform it? It is a nasty problem; how can we not think the way we generally think, use the words we generally use? Philosophers, ranging from Heidegger to Harman to Bernard Stiegler—Latour, as well—make it their point to shape wholly new vocabularies in order to grasp realities or ideas that stay inaccessible to the standard vocabulary, just because of the latter's ingrained presuppositions and "lenses." But the new vocabulary does not become much effective until a critical mass of people start using it, or parts of it. And this is usually something that does not happen, or when it happens, it does so very slowly.

OOO gives us the added advantage that not only can we reflect "within" the theory on what objects/things are. To wit: they "are" *from-out-of* structures—and all structures in this sense are alike. OOO also offers in a meta-theoretical way, if I may put it like that, a sort of fundamental *proviso*, to the extent that the notion of the real object warns us: we can never really know or fully know a thing. This is a subtle nuance. It is on the one hand about being able to realize: all things are broken; our view on them is incomplete. Paradoxically, this realization constitutes again a new sort of "completeness"—*all* things are like that, so we know *that* at least, then. However, on the other hand, thinking through this perspective, one must arrive at a kind of "meta" stance of refusing to know. Of reminding ourselves that we, at base level, do not know what we are dealing with.

I conclude with just a very concise preview of where to go, then, from here. Proceeding from this point on, notwithstanding technology's ontological congruence with all other things, we must—also along the lines of OOO—ask about the particularity of the ontic "technology object." Could we be a bit more specific about technology's characteristics? Put otherwise, how do we delineate it from any other object? We must investigate further how technology at the same time appears and disappears. Elsewhere[10] I attempt to take on this issue by outlining technology in terms of *purpose*, more precisely, of purposive structures, that get deployed at many levels of abstraction. This does not diminish their "weirdness," to put it with Timothy Morton (2018)—to be sure. But it hopefully gives us an extra handle on thinking about them, while navigating *ubiquitous breakdown*. This involves thinking about objects, but *not too much*; thinking about technology, but *not too much*. In other words: being from-out-of-object-oriented, and retaining, as well as embracing, a healthy paradoxical relationship to the "of."

Notes

1. For instance a series of articles on "De Nieuwe Mens" ("The New Human") appeared in the Flemish newspaper *De Morgen*, January 27–February 1, 2018.
2. To begin with, there is Harman's recent *Object-Oriented Ontology: A New Theory of Everything* (2018), which offers at the same time an accessible introduction to OOO and an insight into the latest development of his own theory.
3. Harman: "We expect a philosophy to tell us about the features that belong to *everything*, but we also want philosophy to tell us about the differences between various *kinds* of things. It is my view that all modern philosophies are too quick to start with the second task before performing the first in rigorous fashion" (2018, 55; original emphasis).
4. I draw in this overview especially on Harman's latest summary and development of his theory, in *Object-Oriented Ontology* (2018).
5. For a further exploration of (possible) connections between postphenomenology and OOO, see Van Den Eede (2017).
6. I elaborate this idea more fully in Van Den Eede (2019).
7. Arjen Kleinherenbrink has been exploring possibilities for an object-oriented Sartrean existentialism, exactly on the premise, shared according to him between OOO and existentialism, that existence is irreducible (Kleinherenbrink 2018). Beyond this interesting endeavor, the question further lies open as to how we can nurture in everyday living an attitude in line with this argument; I delve into this in what follows.
8. And in a fundamental way, within the context of OOO, the term "paradox" is obviously misplaced: there is no paradox between substance and relation. It does feel that way, probably because of our background assumptions, but OOO enables us to tolerate the tension.
9. This leads us to a broader discussion about *making* and what that should mean in relation to technology. The fundamental-ontological standpoint of "new objects being created constantly" does not equate, or at least far from fully overlap, with an entrepreneurial call to make, produce, disrupt, and so on. The pleasure in objects coming about and evolving may just as much find a form in *not* (practically) making, and more generally, in not *doing*. But this is a discussion for another place.
10. In the aforementioned monograph titled *The Beauty of Detours*.

References

Achterhuis, Hans. 2001. *American Philosophy of Technology: The Empirical Turn*. Translated by Robert P. Crease. Bloomington: Indiana University Press.

Coeckelbergh, Mark. 2017a. *New Romantic Cyborgs: Romanticism, Information Technology, and the End of the Machine*. Cambridge, MA: MIT Press.

Coeckelbergh, Mark. 2017b. *Using Words and Things: Language and Philosophy of Technology*. New York: Routledge.

Deleuze, Gilles and Félix Guattari. 2004. *A Thousand Plateaus: Capitalism and Schizophrenia*. Translated by Brian Massumi. New edition. London: Continuum.

Ellul, Jacques. 1964. *The Technological Society*. Translated by John Wilkinson. New York: Vintage Books.

Februari, Maxim. 2017. *Klont*. Amsterdam: Prometheus.

Feenberg, Andrew. 2002. *Transforming Technology: A Critical Theory Revisited*. Oxford: Oxford University Press.
Haraway, Donna. 1991. *Simians, Cyborgs, and Women: The Reinvention of Nature*. London: Free Association Books.
Harman, Graham. 2002. *Tool-Being: Heidegger and the Metaphysics of Objects*. Chicago: Open Court.
Harman, Graham. 2009. *Prince of Networks: Bruno Latour and Metaphysics*. Melbourne: re.press.
Harman, Graham. 2010. *Circus Philosophicus*. Winchester: Zero Books.
Harman, Graham. 2011. *The Quadruple Object*. Reprint. Winchester: Zero Books.
Harman, Graham. 2012. *100 Notes—100 Thoughts/100 Notizen—100 Gedanken N°085: Graham Harman, The Third Table/Der Dritte Tisch*. Ostfildern: dOCUMENTA (13), Hatje Cantz.
Harman, Graham. 2014. *Bruno Latour: Reassembling the Political*. London: Pluto Press.
Harman, Graham. 2016. *Immaterialism: Objects and Social Theory*. Cambridge, UK: Polity.
Harman, Graham. 2018. *Object-Oriented Ontology: A New Theory of Everything*. London: Penguin.
Hayler, Matt. 2015. *Challenging the Phenomena of Technology: Embodiment, Expertise, and Evolved Knowledge*. Basingstoke: Palgrave Macmillan.
Ihde, Don. 1990. *Technology and the Lifeworld: From Garden to Earth*. Bloomington: Indiana University Press.
Ihde, Don. 2017. "Foreword: Shadows and the New Media." In *Postphenomenology and Media: Essays on Human–Media–World Relations*, edited by Yoni Van Den Eede, Stacey O'Neal Irwin, and Galit Wellner, vii–xiii. Lanham: Lexington Books.
Kleinherenbrink, A.S. 2018. "Object-Oriented Existentialism." Unpublished paper.
Latour, Bruno. 1993. *We Have Never Been Modern*. Translated by Catherine Porter. Cambridge, MA: Harvard University Press.
Lemmens, Pieter. 2017. "Love and Realism." *Foundations of Science* 22 (2): 305–310. https://doi.org/10.1007/s10699-015-9471-6.
Morton, Timothy. 2018. *Being Ecological*. London: Penguin.
Scharff, Robert C. 2012. "Empirical Technoscience Studies in a Comtean World: Too Much Concreteness?" *Philosophy & Technology* 25 (2): 153–177. https://doi.org/10.1007/s13347-011-0047-2.
Smith, Dominic. 2015a. "Rewriting the Constitution: A Critique of 'Postphenomenology.'" *Philosophy & Technology* 28 (4): 533–551. https://doi.org/10.1007/s13347-014-0175-6.
Smith, Dominic. 2015b. "The Internet as Idea: For a Transcendental Philosophy of Technology." *Techné: Research in Philosophy and Technology* 19 (3): 381–410. https://doi.org/10.5840/techne2015121140.
Smith, Dominic. 2018. *Exceptional Technologies: A Continental Philosophy of Technology*. London: Bloomsbury.
Sutherland, Thomas. 2013. "Liquid Networks and the Metaphysics of Flux: Ontologies of Flow in an Age of Speed and Mobility." *Theory, Culture & Society* 30 (5): 3–23. https://doi.org/10.1177/0263276412469670.
Van Den Eede, Yoni. 2016. "The (Im)Possible Grasp of Networked Realities: Disclosing Gregory Bateson's Work for the Study of Technology." *Human Studies* 39 (4): 601–620. https://doi.org/10.1007/s10746-016-9400-x.
Van Den Eede, Yoni. 2017. "The Mediumness of World: A Love Triangle of Postphenomenology, Media Ecology, and Object-Oriented Philosophy." In *Postphenomenology and Media: Essays on Human–Media–World Relations*, edited by Yoni Van Den Eede, Stacey O'Neal Irwin, and Galit Wellner, 229–250. Lanham: Lexington Books.

Van Den Eede, Yoni. 2019. *The Beauty of Detours: A Batesonian Philosophy of Technology*. Albany: SUNY Press.
Van Den Eede, Yoni, Gert Goeminne, and Marc Van den Bossche. 2017. "The Art of Living with Technology: Turning over Philosophy of Technology's Empirical Turn." *Foundations of Science* 22 (2): 235–246. https://doi.org/10.1007/s10699-015-9472-5.
Verbeek, Peter-Paul. 2005. *What Things Do: Philosophical Reflections on Technology, Agency, and Design*. Translated by Robert P. Crease. University Park (PA): The Pennsylvania State University Press.
Verbeek, Peter-Paul. 2015. "Toward a Theory of Technological Mediation: A Program for Postphenomenological Research." In *Technoscience and Postphenomenology: The Manhattan Papers*, edited by Jan Kyrre Berg O. Friis and Robert P. Crease, 189–204. Lanham: Lexington Books.

11 THE DISAPPEARING ACTS OF THE MORSE THINGS: A DESIGN INQUIRY INTO THE WITHDRAWAL OF THINGS

Ron Wakkary, Sabrina Hauser, and Doenja Oogjes

I was also thinking that the home that you live in sort of contains these lost things. It contains a lot of lost socks, they are somewhere. The bowls are in the same way sort of contained.

—ELLA COMPARING THE MORSE THINGS TO LOST SOCKS

INTRODUCTION

While we may not fully understand the things we live with, we are able to coexist and form relations with them. We may not know where our lost socks go, and neither do we fully understand the ways in which virtual agents like Alexa™ or Siri™ know which music we like best. Things withdraw from our human understanding and perception, and that subsequently contributes to a gap between things and us (Wakkary, Hauser, and Oogjes 2018). In other words, much of what makes up our experience of and with things is beyond our grasp. We believe it is relevant to look closer into this gap.

We have dealt with this notion of withdrawal in our generative and empirical design research in which we inquire into human-technology relations through the crafting and studying of research artifacts. In our design research, we have investigated the ontological gap between humans and things in thing-oriented inquiries (Wakkary et al. 2017), the concept of "displacement" in which the relations with things are often obscured and a matter of an incompleteness (Wakkary,

Hauser, and Oogjes 2018),[1] the challenge of representing the hybrid relations of things in design from a postphenomenological perspective (Oogjes and Wakkary 2017), lived experiences of philosophers revealing aspects of background relations and relativistic views of a novel digital research product (Wakkary et al. 2018), meta-reflective analyses of challenges we have faced in doing postphenomenology through design research (Hauser, Wakkary, et al. 2018), and a critical analysis of how speculative design research in Human-Computer Interaction (HCI) can be seen as holding postphenomenological commitments (Hauser, Oogjes, et al. 2018).

Several methodological innovations have helped us better understand human-technology relations. In this chapter, we introduce *material speculation* and *co-speculation* as a way to come closer to the ways things withdraw from us. We do this through one of our design research cases—the *Morse Things*.

The Morse Things study investigates the nature of living with everyday things that are networked together and communicate with each other in what is commonly understood as the Internet of Things (IoT). The study asks: *What might be revealed in the relations we have with technologies through a thing-oriented approach to IoT?*

In the Morse Things study, we designed and fabricated six sets of networked ceramic bowls and cups (Figure 11.1) to be given to professional designers and design researchers—to live with for several weeks and to inquire with us on their experiences. Over time, the conversation of the Morse Things and their degree of connectedness on the network can evolve in degrees of "awareness" from being alone, to being a pair of things, to being a group of things, to being part of a larger network of things. The Morse Things will send and receive messages to and from other Morse Things in its set at random intervals during the day, at least once

FIGURE 11.1 A set of Morse Things.

every eight hours. The messages sent by each Morse Thing are in Morse code and simultaneously expressed sonically and broadcasted on Twitter. The Morse Things can be used like any other bowl or cup for eating, drinking, and containing items.

In this chapter, we focus on our Morse Things study since the matter of withdrawal emerges as a gap in the intelligibility of the relations between us and things from the perspective of both those who live with the Morse Things and us as design researchers who designed the Morse Things. Besides having an incomplete picture of the things around us, we may still enter into relationships with them. The design research case of Morse Things reveals three themes of these relationships: (1) searching for humanness, (2) thing-centeredness, and (3) tensions in making sense of the gap between things and us that we extend into a discussion of the withdrawal of things. At the conclusion of the chapter, we take a step back to explore the potential partnership of philosophy of technology and design-oriented Human-Computer Interaction, our main research field.

Background

In the following sections, we provide conceptual background for and related work to our generative and empirical design research efforts with the Morse Things. We discuss related philosophies of technology, thing-oriented research in HCI, and our understanding of the notion of withdrawal.

Philosophies of Technology

In our work with the Morse Things project, we draw on philosophy of technologies, mainly postphenomenology and object-oriented philosophy.

Briefly, postphenomenology, as argued by Don Ihde and Peter-Paul Verbeek (Ihde 1993; Verbeek 2005), understands technologies as mediators of human experiences and practices rather than functional and instrumental objects (Verbeek 2005; Rosenberger and Verbeek 2015). In a postphenomenological relationship between humans and technological artifacts, each mutually shapes the other through mediations that form the human subjectivity and objectivity of any given situation, and this gives rise to a hybrid relationship between us and things. Design is central to and bound up in a postphenomenological understanding of the world since digital technologies do not come to us in a "raw" form but in a form that is *designed*.

In this respect, designed digital artifacts, or in our case *things*, manifest technologies and directly influence the mediation of our experiences and practices. Beyond postphenomenology, recent philosophical thinking like object-oriented philosophy (Harman 2010) has adopted more radical thing-centered approaches that advance the position that things and artifacts bear knowledge in distinct and complex ways. While there are important differences between the various epistemological commitments, emerging theoretical notions—such as Ian Bogost's *carpentry*, the construction of artifacts that do philosophy (Bogost 2012), Graham Harman's speculative realism that critiques anthropocentrism and the undermining of objects in philosophy (Harman 2010), and Davis Baird's *thing knowledge* in which artifacts embody and carry knowledge prior to our ability to theorize or reason through language (Baird 2004)—offer intriguing perspectives that can be seen both as critical and as generative mechanisms for post-human approaches to design.

Drawing on the philosophies and philosophers of technology such as Ihde, Verbeek, Albert Borgmann, and Bruno Latour (Borgmann 1987; Ihde 1990; Latour 1992; Verbeek 2005, 2011) is not new to design and HCI research (see, for example, Fallman 2011). Recent research in HCI like that of Odom et al. (2009) describes attachment as a key factor in human-technology relations for future design implementations. Pierce and Paulos (Pierce 2009; Pierce and Paulos 2011, 2013) aim to describe the materializing of technologies and its implications from the material awareness of everyday things to embodied relations within technologies. Relatedly, Tromp et al. (Tromp, Hekkert, and Verbeek 2011) reflect on the social consequences of mediated relations and argue that designers should make more informed decisions to design for socially responsible behavior. Our investigation similarly focuses on the role of the thing yet moves beyond materials or embodied interaction or moralizing behaviors to articulate the complex and ambiguous relationships that form between things and us.

Thing Perspectives in HCI

Drawing on the philosophies of technology and object-oriented philosophy, we can see that the notion of a *thing* is neither in reference to technologies nor simply artifacts in the physical sense. Things can be seen as nonhuman technological entities and artifactual entities often bound together that are conditioned by humans and in turn shape what it means to be human. As such, they have a central importance to HCI as computational artifacts, systems, and processes that can be referred to as things.

Our work is mainly situated in design-oriented HCI research. In what follows we briefly discuss related works from this field.

Crabtree and Tolmie (2016) explore a "day in the life" of things through analyzing a series of mundane interactions within a household. It is an investigation of the challenges to the design of IoT things. This ethnomethodological study portrays a perspective of things in order to uncover the underlying "machinery" of interactions that tends to fade into the background yet governs the meaning of things in our everyday lives. This research provides a thorough but explicitly human view of things from third-person human perspectives.

Related research has taken more literal approaches to the notion of a thing-perspective. For example, the PetCam (Keeney 2014), BinCam (Comber et al. 2013), and FridgeCam (Ganglbauer, Fitzpatrick, and Comber 2013) utilize small cameras embedded or attached to objects (and pets) to provide a visual perspective that is quite literally from the perspective of things. On the surface, these approaches appear to provide viewpoints unfamiliar to humans, and in this light, they reveal new insights and observations. For example, Giaccardi et al. (2016) explicitly introduce things as co-ethnographers in a study that attaches cameras to household items like a kettle and cups to log a visual perspective on human actions and routines from the vantage point of artifacts. The Long Living Chair (Pschetz and Banks 2013), by contrast, does not set out to observe human actions; rather, it embodies the relation to humans by detailing the day it was manufactured (i.e., its human age) and records and displays how many times it has been used. In contrast to our post-human approach, these works adopt a thing-centered perspective of concerns that are in essence human-centered. The purpose and role of these approaches are focused on human actions and activities whether to observe people or to embody and record human interactions.

Other work aims for a more radical thing-oriented approach. For example, Trojan Boxes (Davoli and Redström 2014) utilizes embedded cameras but turns its view to the lived experience of things that are mostly nonhuman encounters, revealing a world most people have not experienced. The Trojan boxes are mail parcels with a tilt-triggered camera inside to document the various stages in our global delivery system of goods. However, here these works adopt an anthropomorphic standpoint from which the human privileging of visual sight is assumed as a way of being for things. This view overlooks the alien and inaccessible aspects of the withdrawal of things with respect to human perception and reasoning.

Withdrawal

As we discussed earlier, understanding experience from the perspective of things is fraught with intelligibility issues that make this a difficult task for people. In response to the alien nature of things, there is a tendency to relate to these nonhumans as surrogate humans, that is, to anthropomorphize, as discussed above. Latour argues that anthropomorphism is not in itself a problem and in fact can productively frame things on a continuum of delegation between human and nonhuman elements in which artifacts take on human functions and vice versa (Latour 1987). However, complicating this symmetry between humans and nonhumans, nonhuman perspectives can be said to "withdraw" from human understanding into a nonhuman world that humans can neither fully comprehend nor articulate (Verbeek 2005; Bogost 2012). In addition, nonhuman worlds are formed in a configuration of materials and performances rather than language (Baird 2004; Bennett 2009).

We should clarify that our use of the term "withdrawal" borrows from object-oriented philosophy. As such, it is different than the notion of technology withdrawing from our perception to be a transparent extension of our own bodily senses. This is often characterized as the "ready-to-hand" experience of the hammer as described by Heidegger (1962). Ihde critiques this notion as only one of many possible relations with technology. Further, he argues it is a misguided and contradictory desire to have technology transform our human capabilities without our awareness of its presence, overlooking the mediating effect that a thing simultaneously amplifies certain bodily aspects while reducing others (Ihde 1990, 74–75).

In our use of the term, withdrawal underlines the difference in embodiment between us and things. However, here we consider this difference and withdrawal within the confines of hybridity between humans and technology or within the subject-object schema of postphenomenology. That is, while humans and nonhumans are "interwoven and mutually constituted," it is "meaningful to make a distinction between someone who experiences and something that is experienced." Further, following Verbeek, to draw such a distinction between humans and nonhumans does not suggest a separation of subject and object but rather affirms that in human experience, the difference or perceived withdrawal of things is a "vivid reality" (Verbeek 2005, 166–176). In this sense, we assume that if a thing is intelligible or experiential, even in a limited form, it is of a hybrid relation.

In the Morse Things study, we highlight themes that explore withdrawal as a tension between humanness and thing-centeredness in the relations of humans and things. We discuss these themes later in this chapter as qualities of incompleteness and unknowing. In the next section, we will elaborate on our empirical design strategies that enabled our investigation into withdrawal: *material speculation* and *co-speculation*.

Our Empirical Design Research Strategies

In our work, we craft and study what we refer to as *research products* (Odom et al. 2016), which are artifacts designed to drive a research inquiry, have a quality of finish so people engage it as it is rather than what it might become, fit in everyday settings and be lived with over time, and be independent such that it operates effectively when deployed in the field for an extended duration.

With and through our research products, we are able to inquire empirically into human-technology relations, for example, through deploying these artifacts over time with others to live with, and thereby engaging in first- and third-person perspectives. Within our approach lie specific methodological innovations we have developed that are particularly productive in dealing with the gap between things and us and the withdrawal of things: material speculation and co-speculation. Material speculation and co-speculation move our design-philosophy inquiries past retrospective reflections of existing things to generative and material understanding of the research concerns, offering a new empirical approach for design research and philosophies of technology.

We will describe these strategies and how we deployed them in the Morse Things project.

Material Speculation

In our design research investigations into relations between humans and things we developed a methodology we call material speculations (Wakkary et al. 2015b), in which we design artifacts that are crafted to embody research questions or propositions to be lived with. We refer to these artifacts as *counterfactual* because they are designed against current norms of design artifacts as a way to elicit or make visible new phenomena that we as the designers cannot fully anticipate or know in advance. Our inquiries emerge from the lived-with experience and observed existence of the counterfactual artifacts.

The IoT is concerned with the design of internet-connected interactive artifacts enabled to collect and exchange data. Morse Things are counterfactual IoT artifacts in that their digital capabilities are at the service of things rather than people. Their functionality in the service of humans is of an everyday nature that already exists in homes today (i.e., the functionality of a bowl in a domestic setting), shifting the question from what they *do* to how they *are* in our homes. The Morse Things embody the proposition that our relationships with internet-enabled things are a matter of negotiation over time rather than predefined or prescribed as a service or functionality.

We purposely limited the communications of the cups and bowls to communicating with each other and to affirming their individual or group existences on a network. The aim here was to foreground a thing-oriented approach. In a sense, we designed the Morse Things to ask the question: *What is it like to be a thing on a network*? We exaggerated the thing-oriented approach by designing computational technological functions of the Morse Things to the exclusion of people, to have the objects computationally exist in their own world, so to speak, relatively independent of human action. While they can be used as any other bowl or cup for eating, drinking, and containing items, this use does not impact their communication or "awareness"; nor are these interactions with the Morse Things sensed or data logged. We constrained the computational technologies to be solely at the service of the Morse Things. These strategies are combined as a way to both acknowledge and inquire upon the gap between things and us.

We chose to design ceramic bowls and cups because we wanted our Morse Things to readily fit with and be accepted like any other household object in order to perform the inquiry of a material speculation. The combination of technological and nontechnological identities within the Morse Things underscored our use of *defamiliarization* as a technique common to speculative design (see, for example, Bleeker 2009). To defamiliarize is to make the familiar strange as a way to call into question the usual interpretations of everyday or known things.[2]

Enforcing the gap between technological things and us was important; however, it was equally important to design links and reminders of the potential relations between the things and us despite this gap. The physical form of the

Morse Things was aimed to keep present the idea that the bowls and cups were also technological objects, although the electronics were hidden. The form of the bowls and cups protrude in odd shapes (unlike other ceramic bowls), revealing where the electronics fit between inner and outer ceramic shells (see Figure 11.1). Similarly, we chose to make the communication between the Morse Things potentially intelligible to people by having them in Morse code, sonically expressing the messages, and translating the messages on Twitter. We chose Morse code as potentially familiar yet an outmoded form of communication that is for human communication yet designed for the mechanical and electronic properties of nonhuman things. Twitter was chosen as the Morse Things' communication platform because it was easily integrated into our system, enabled participants to monitor the communications easily, and is reminiscent of other IoT things on Twitter (see, for example, @mytoaster ("Mytoaster (@ mytoaster) | Twitter" n.d.)).

To summarize, material speculation is the design of a counterfactual artifact that is experienced and lived with on an everyday basis over time as a way to ask certain types of research questions (Wakkary et al. 2015a). A counterfactual artifact is a realized functioning product or system that intentionally contradicts what would normally be considered logical to create given the norms of design and design products. This countering of norms opens the possibilities to empirically investigate multiple alternative existences (or what-ifs) as lived-with realities of the counterfactual artifacts.

Co-speculation

Co-speculation is the recruiting and participation of study participants who are well positioned to actively and knowingly speculate with us in our inquiry in ways that we cannot alone. A key motivation in this approach is the desire to diversify and deepen the reflective competences and perspectives to better describe and investigate the nuanced and challenging notions of human-technology hybridity, especially in light of the withdrawal of things.

For example, in a different study we recruited trained philosophers who have the competences (e.g., critical thinking, ethical training, and philosophical vocabulary) to help us speculate, reveal, and describe human-technology relations with a counterfactual artifact we made known as the Tilting Bowl (Wakkary et al. 2018). Philosophers both lived with and actively reflected on the Tilting Bowl for a lengthy period of time in their own homes.[3] In the Tilting Bowl study, we aimed to address the philosophical aspects of our concerns directly through the use of co-speculation.

In the case of the Morse Things, we recruited professional designers and design researchers who on some level had experience with aspects of IoT. The designers, design researchers, and their families were asked to live with the Morse Things for six weeks and to document the experience (see Figure 11.2) in response to a question and task we provided: *Describe what it is like to live with the Morse Things from the perspective of the Morse Things?* And *Design an artifact, system, or service to coexist with the Morse Things.* It was left to the participants as to the form of their response and the form of the concepts they generated.

At the end of the deployment, we organized a workshop with all of the designers and some of their household members to discuss the role of the Morse Things and ultimately the idea of the distinction between things and humans. During the workshop, which lasted approximately six hours, the participants were asked to first present individually their experiences of living with the Morse Things and give special attention to the deployment question about the experience of the Morse Things from the perspective of the things. In a second round of presentations, participants presented their individual concepts of things designed for the Morse Things. Then, participants were divided into two groups and engaged in a group activity of designing things for the Morse Things as a group. Throughout these workshop activities, we engaged in in-depth discussion with our designer participants, which was as lively as it was informative.

To summarize, co-speculation is a form of collaborative inquiry with expert study participants. Co-speculation occurs through and while living with a research product that drives the speculation. This co-delegated approach to

FIGURE 11.2 Participant photos of Morse Things in their homes.

FIGURE 11.3 Each Morse Things set included a large bowl, medium bowl, a cup plus a Wi-Fi hub and a set of instructions.

empirical research allows us to diversify and deepen our investigation of complex and subtle aspects of human-technology relations.

Morse Things Study and Emerging Themes

Our study included six designers and their households. For each household, we provided a set of Morse Things. Each set included a large bowl, medium bowl, a cup plus a Wi-Fi hub and a set of instructions (see Figure 11.3). The box also contained an information sheet specifying the messages that Morse Things send, how they translate, and when they occur and how to access the messages on Twitter. While living with the Morse Things, our participants gathered photos, video, and text entries from diaries, as well as sketches of concepts.

In analyzing the data from the deployments and workshop three main themes emerged: (1) searching for humanness, (2) thing-centeredness, and (3) tensions in making sense of the gap between things and us. Our analysis is organized to reflect the shift in thinking of our participants from viewing things as human-related to then thing-related and lastly the tension of holding these two positions simultaneously.

Searching for Humanness

Projecting Human Qualities onto the Morse Things

Participants projected human emotions and experiences on the Morse Things. For example, Hannah, a senior designer and digital strategist, translated the messages of the Morse Things into more human language. She also considered the "emotional life" of Morse Things like feeling lonely, frustrated, bored, forgetful, restless, and ignored. Olivia, a professor in interaction design, described the reactions of the Morse Things to events in their "lives." For example, she thought the Morse Things would be happy with their new home, and as the Morse Things made sounds when she and Noah entered the house, she imagined them to be happy to see them: "They were here and they spoke a little bit and then we went out for dinner… we came back… and as we entered the door, someone, one of them was like bipbipbip, and I was like, Oh! He's so happy to see us!"

Comparing Morse Things to Family Members and Pets

Along the same line of analysis, the Morse Things were compared to humans and animals. For example, Olivia described the Morse Things as a family that stayed together as a set. The Morse Things were compared to children in different instances. Noah, Olivia's partner and a landscape designer, described them as young children that were learning and evolving. Ethan, who is also a professor in design, participated in the study with his wife Emily, a media professional. In their presentation during the workshop, Ethan and Emily emphasized that their son Edwin could most easily relate to the Morse Things because "he is already in that space, making Lego and doing things." Most of Spencer's concepts were inspired by children's toys motivated by a desire to maintain the abstract and playful aspects of the Morse Things. Spencer is a professional interaction designer. Both Ella, a professor in design and textiles, and Emily compared the Morse Things to teenagers, as well as to cats thinking of them as going their own way: "I think what they do is make us aware that there's other things going on that we have no idea about, like with the teenagers.… I don't know what the cat is doing when I'm sleeping or what my kids are doing."

Morse Things Seen as Being Aware of Humans

In responding to our request to document the experiences of the Morse Things from the perspective of the things, participants described the Morse Things as thinking of and being aware of the people in the house. Spencer describes conversations between the bowls and their thoughts of humans, which in his account they call "strange giants": "I get used the most in the morning. But, not the way I expect. The big one usually puts some sugar and milk in me before the warm brown liquid. Which is strange, because it does it the other way with those stupid mugs."

Ella also describes, in a short-sentenced robotic way, the Morse Things' awareness of the people in the house: "Human number four only heard us once…; Human number two has been waiting for a…; Human number one is remaining objective…; Human number three can't count our attempts to connect."

Control and Feedback Modalities for Humans

In their concepts, the designers indicated a desire to better connect with the Morse Things. For example, when the Morse Things would make a sound, Spencer tried to keep track of which Morse Thing was saying what but always found himself too late to distinguish and locate the sound: *"Maybe an LED in the rim or something, which got brighter every time there was some activity, and then faded out over time to give you some indication of how active they had been relative to each other."* Participants' concepts also included giving the cups and bowls personalities which could reflect in the aesthetics of interaction, for example, sound and light. Hannah was specifically looking for added feedback modalities motivated from a more empathetic standpoint, in which she mainly wanted to make sure the communicating cups and bowls were finding each other: "I thought that if one was seeking, another could vibrate." This empathy with the perceived struggle of the Morse Things connecting to each other was also reflected in the concept of another participant, who envisioned a device that would manage the timing of their messages, enabling the Morse Things to be "awake" and communicating at the same time.

Connecting to Human Practices

The desire to connect the Morse Things to human practices emerged too: "If they could detect us through motion, or maybe just by touch, if we pick them up and all of a sudden they vibrate or they spoke or they can feel us touching them." Beyond direct interaction, participants were envisioning ways in which the Morse Things could work themselves into daily human routines (see Figure 11.4). Ethan's concept positioned the Morse Things as use-logs that would remind you of your last activity with them twenty-four hours later, to recall or even dictate routines, where Hannah's concept argued for having the Morse things as melodic, harmonious companions, providing moments of reflection in everyday mundane things: "They would wake up with us, as we are starting our day. So if I walk past a cup, and maybe a coffee maker, it would just chime as I walk by it, and as I do more it becomes a bit more musical."

Thing-centeredness

Engaging with the Morse Things' Language

While the Morse Things were often approached from a human-centered perspective, on many occasions a more thing-oriented interpretation came through. Participants

Building micro-moments for meditation

What if we could wake up objects in our home as we start our day?

What if our objects could work in harmony with one another?

What if our objects could build calm in our busy lives?

FIGURE 11.4 Hannah's concept of Morse Things as part of daily routines.

felt like they themselves needed to learn more about being a thing to understand the Morse Things. Noah mentioned that it might take more time to understand the Morse Things: "Maybe it's a process that takes a longer time to really see. Because, like everything takes time to evolve and change, and maybe the speed of that discussion is like that. Maybe not a computer speed evolution, more like a human evolution. More slow basically, versus technology going really fast."

Noah and Olivia also talked about learning Morse code to understand the Morse Things' conversation. Sandra, Toby's partner, wanted to tell them apart: "I continue to keep trying to grab the bowls while they are 'tweeting.' I don't know why I'm doing this, because I can just wait and check Twitter to see which bowl it was... guess I feel like I might be able to learn if they have different sounds? Maybe I'll be able to tell them apart eventually."

Things with Other Things

In our participants' concepts, the Morse Things were often connected to other objects in their environment. Concepts included the Morse Things as eavesdropping bowls that listen in to you and your devices, and the Morse Things as silent ethnographers informing an electronic tablecloth that keeps track of the activity in a coffee shop. Other concepts looked at possibilities of how the Morse Things could include other things in the house on their network; for example, Olivia proposed the idea that the Morse Things could "hack" into other things in the home, like televisions, to join their network (see Figure 11.5).

Comparing to Other Things

In understanding the Morse Things, our participants compared and related them to other existing things. Ella compared them to lost socks: "I was also thinking that the home that you live in sort of contains these lost things. It contains a lot of

FIGURE 11.5 Olivia imagined the Morse Things could hack into other things to join their network.

lost socks, they are somewhere. The bowls are in the same way sort of contained" (see Figure 11.6). She continued this comparison in explaining why the bowls were useful and useless to her at the same time:

> [T]hose are things that are in our homes and they are just, they're there. I suppose that's why I went to that space because although we think of bowls functioning in a particular way, we put things in it, as another type of entity that has a digital life, it wasn't functional. It was like the lost socks. We have an object that actually functions, physically, but we have an object that is just there.

Noah did not see the Morse Things as different from other cups and bowls and compared them to the Nest thermostat:

> To me, there's technology in it, but I look at it as a thermostat. I don't see it as being a new everyday complex thing. I look at that thing and I look at the Nest thing, and it's the same thing for me. So the same with the bowl, I look at that bowl and I look at the other bowl in the cupboard: same thing.

He also compares the Morse Things' conversation to playing a compact disc (CD) or running a script on which he has no impact: "It feels like it's a CD that plays on a loop. It plays that and it just keeps going."

Toby, a professional interaction designer, introduces the idea of the Morse Things being a new class of object and positions them between a digital product and a puzzle or a painting:

> It's kind of a new class of object. I was thinking is there any non-digital object in my house that is actively disrupting the environment for its own pleasure,… just to please itself. And there isn't really. Sometimes you'll have like a puzzle or an object you misread, like maybe you got a painting and you still haven't really figured out what it is, what it means. But that's different, because it's

FIGURE 11.6 Ella's concept for finding and containing things.

passively there, and you choose when to engage with it and you get to make some meaning. You're not really trying to put it to any purpose, whereas the bowl… sometimes you are trying to put it to purpose and then it just interrupts you and is like hey, figure me out.

Tensions in Making Sense of the Gap between Things and Us

The tensions in making sense of what the Morse Things are and what role they could play in everyday life were very present throughout the study. While

Olivia "loved imagining" that the Morse Things talked and cared for her and her partner, she realized that "that's not what they're saying at all, and they don't care about us at all." Emily described a situation in which she was not home but saw on Twitter that the bowls had been making sounds. She continued with her comparison to cats and teenagers: "And there is something kind of nice about not knowing,… but with a bowl, that's where it sort of gets strange." Ella recognized this friction and attributed it to the fact that the bowls are not conscious beings:

> If my kids are going out and they have a relation and they are talking with each other—I don't know what's going on but I know they're doing it—I at least know that they are conscious and aware of it. With these bowls I know that they are not. That's why it seems like… why are we doing that for them if they are not conscious or something going on.

Continuing with his comparison of the Morse Things to a CD or script, Noah wondered whether the experience of the Morse Things is actually ours or the designers or researchers who made and programmed them. This friction in what the Morse Things are or should be continued in participants' concepts. For example, Spencer added explicit functionality to the Morse Things, in modifying them so they would function as both a Wi-Fi repeater and a plant-watering reminder system (see Figure 11.7), while Spencer avoided making it useful, as he wanted to keep thinking of them as an abstract, playful thing, rather than something utilitarian. Hannah mentioned that she did not need the Morse Things to have more functions: "It doesn't need to have a specific use, I like that they are just there and kind of in their own world and speaking in their own language, and sometimes my interactions with them impact them."

Should Things Only Exist for Us?

In the discussion at the end of the workshop, the participants were divided in their opinions on whether the Morse Things should exist for us, as Ella argued, "If it can talk, allow it to talk to us. If it's communicating then we want to have a conversation," or whether they should exist on their own as Spencer says, "That's why I like the idea of something else, let them be themselves. Other stuff is going on that we're just totally unaware of and it doesn't matter."

Discussion

We based our investigations on the argument that the lives of things are neither fully perceivable nor comprehensible to human understanding. The discussion that follows represents an answer to the question we posed earlier in the chapter: *What might be revealed in the relations we have with technologies through a thing-oriented*

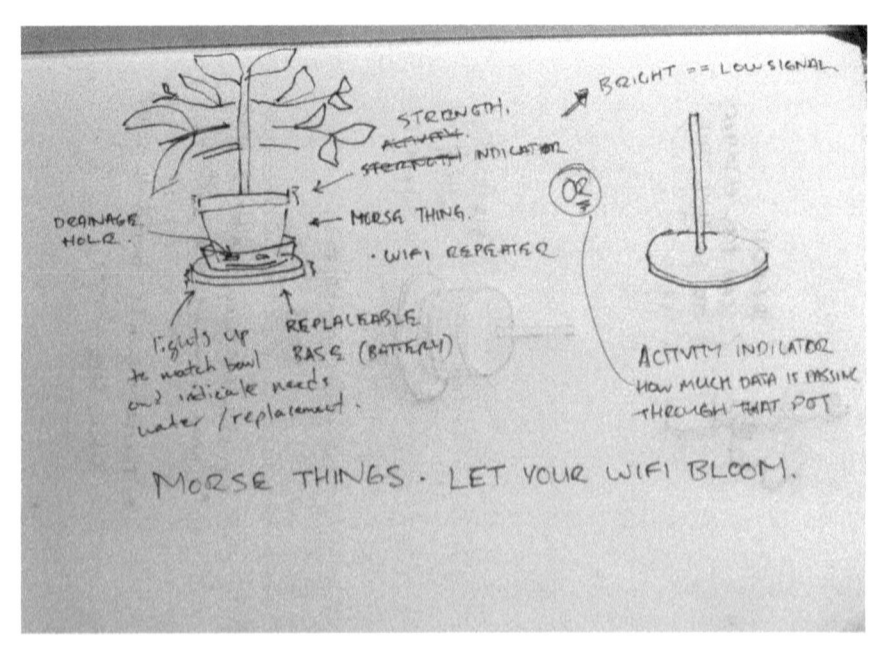

FIGURE 11.7 Spencer's concept of having the Morse Things as a Wi-Fi repeater and watering system.

approach to IoT? Aspects of the answer include the withdrawal of things from our human understanding and perception that contributes to the gap between things and us, the ability to form attachments with things despite withdrawal, and lastly the understanding of unknowing and incompleteness in the relations between us and things.

Withdrawal of Things from Human Understanding

As we discussed, understanding experience from the perspective of things is a difficult task for humans. Philosophically speaking, nonhuman perspectives "withdraw" from human understanding into a nonhuman world that we can neither fully comprehend nor articulate (Verbeek 2005; Bogost 2012). This challenge was confirmed in our study as our designer participants readily described the Morse Things as having human qualities like an "emotional life" (Hannah) or belonging to a family (Olivia). In imagining the perspective of the Morse Things, participants gave them language and forms of agency. For example, Spencer saw the Morse Things as human-like characters that perceived humans as "strange giants." Ella described the Morse Things similar to how one might describe robots that express themselves like humans but through logic and without emotions.

It is important, though, that these interpretations were not resolute and our participants knowingly held contradictory views of the Morse Things. These knowing contradictions acknowledge the difference between what we imagine of things and how they actually exist. Referring again to comments made by Olivia (see Tensions in making sense of the gap between things and us), she said that despite the fact that she "loved imagining" that the Morse Things talked and cared about her, she knew they did not care at all about her.

Our study gave details on experiencing the gap between things and us but also revealed the more nuanced idea of how things withdraw from our understanding, as has been expressed by Verbeek (2005), Bogost (2012), and others. It is important that while much of the experience of things is beyond our grasp, this perspective is not entirely invisible to us. Rather we establish many commonalities and reliable interactions that form the foundations for the fundamental and ubiquitous relations we have with things. This relationship to things in the context of their withdrawal emerged clearly in our study. For example, we reported on how Ethan and Emily believed Edwin, their four-year-old son, could best relate to the Morse Things since he spent his day playing in an imaginary world of things. Ella and Emily throughout our workshop compared the Morse Things to pets and teenagers signaling familiar relationships that at times are very unfamiliar if not inaccessible to us.

This semi-independence of things relates to Verbeek's discussion of the intentionality of artifacts in which things play an active role in shaping human actions and interpretations. Things can also be said to be capable of original intentionality in which the mediations that arise were not the intentions of the users or designers, as is the case with the Morse Things. Thing intentionality in this respect is a unique form that is material and dependent on human intentionality, resulting in a hybridity that is "a complex blend of humanity and technology" (Verbeek 2009, 272–273).

Attachment with Things We Don't Understand

Despite this gap between things and us, it was evident in our study that participants formed attachments with the Morse Things. After the initial curiosity subsided, the Morse Things were momentarily forgotten or ignored but later became part of the daily lives of the homes. This was clear in the reports and images sent to us during the deployment. In addition, the Morse Things' messages were routinely checked on Twitter, and participants spoke about taking care of their set. In one incident, two sets were accidentally swapped during a maintenance check, and both households immediately demanded back their own set (each set is a unique combination of colors). Lastly, at the end of the deployment, nearly all participants wanted to keep the Morse Things. This attachment with the Morse Things was not necessarily a foregrounded experience but rather one in which the Morse Things faded into the background of everyday living to on occasion surface in ways that caused reflection, new considerations, and even pleasure and comfort. Toby's comment on his experience of "rediscovering" a Morse Thing speaks well to a type of attachment that was common in the study: "Finally heard a bowl! It's been a week. I didn't expect that I would be as surprised or excited as I ended up being. Had a pretty good rhythm to it. Dah-do-dah-do-do-dah-dah-do-dah-dah … or something like that."

The attachment to the Morse Things was also described in ways that acknowledged that the relationship was with things that could not be fully understood—that the relationship could be more thing-centered than human-centered. Noah commented on how it may take time to fully develop a relationship with things that in the case of the Morse Things would be at a slow evolutionary speed rather than computer speed, observing the difference between things, humans, and computers. Noah and Olivia also discussed learning Morse code to better understand the Morse Things from their perspective. Lastly, Ella elaborated that the attachment was a matter of things that function on some level (being a bowl), but digitally they are "just there" in our home.

Incompleteness and Unknowing in the Relations with Things

We propose that the inherent contradictions, the counterfactual nature of the Morse Things, reveal qualities of incompleteness and unknowing in our relations with things. The Morse Things are not typical everyday artifacts, like other bowls or cups in the home, but you can just as easily use them like a typical bowl or cup. They are also not digital devices, like mobile phones or smart thermostats; however, they are internet enabled and connected.

Toby explicitly focused on the novelty of the Morse Things. He considered them to be a "new class of object" that he compared to artifacts like a painting. One does not fully understand a painting; yet, one forms a special relationship that spurs ongoing reflections and interpretations despite knowing these can never be resolved. However, unlike paintings, Morse Things have "autonomy" and "interrupt" or emerge into our lives on their own accord. This in our view reveals the productively incomplete nature of relations to things rather than the newness of the thing itself. In such a relation, we become ambiguously attached to things in our daily life that are in many respects independent of our actions and desires. This notion of thing relations opens IoT approaches to consider IoT things that form attachments with people through qualities other than human-centered functionality or explicit services. Ambiguity in this sense is a resource (Gaver, Beaver, and Benford 2003) that adds dimensionality and complexity that is more commensurate with human-technology relations; Verbeek reminds us that technologies are mediators of human experience and practices rather than functional and instrumental objects (Verbeek 2005).

Ambiguity or ambivalence in relation to things is important to recognize. In keeping with the complexity of our relations to things, it is not clear why these relations should be resolved or satisfying in order to maintain an attachment. Maintaining relations while not knowing is critical to future possibilities and alternate meanings that ultimately sustain our relationships with things— contrary to the functionalism that underpins much consumerism. In our study, this ambivalence with the attachment to Morse Things emerged. Ella, in her comparison of the Morse Things with teenagers, makes the point that they are "conscious and aware" as opposed to the Morse Things that "are not conscious," and yet we are concerned for them in ways similar to our complex relationships with teenagers.

In existing IoT systems we see glimpses of nonhuman agency and thing-centeredness in systems that automate updates of mobile operating systems and applications or service and maintenance notifications of appliances and automobiles. However, these are human-centered in their orientation, focused on automating human tasks. An overlooked example but one that is more relevant may be the accidental relationships that form between digital things and other things, like audio speakers that unintentionally convert nearby cellular radio transmissions of mobile phones into sound that can reveal incoming data or phone calls. This unintentional thing-to-thing interaction reveals independent but intelligible thing-centered interactions. In other research we have discussed the idea of *ensembles* (Odom and Wakkary 2015; Wakkary, Desjardins, and Hauser 2015) in which over time things configure into relations *seemingly* on their own. Examples of this include complex arrangement of objects and furniture in your apartment or home, or how keys always find themselves in a bowl on a table near the front door. Human actions co-mingle with nonhuman qualities to form ensembles that demonstrate difficulty to describe relations as the Morse Things suggest.

A Dialogue between Philosophy of Technology and Design Research

At the outset of the chapter, we said we would like to further open the dialogue with postphenomenology in ways that are mutually beneficial to HCI or design research. In bringing HCI together with philosophy of technology, as we do in our research, we afford ourselves an empirically driven philosophical account of living with technological things. Our philosophical approach brings to HCI a framing and set of concepts not typically considered in understanding how to design with technologies. These include relational ontology, human-technology relations (embodied, hermeneutic, alterity, and background), and multistability for starters. More recent research has extended human-technology relations to include a set of cyborg relations that account for body implants, home automation, and augmented reality (Verbeek 2008) that are obviously related to HCI.

HCI, in turn, brings to postphenomenology the opportunity to proactively design a technological artifact and tailor it to an inquiry as we did with the Morse Things (also see: Hauser, Oogjes, et al. 2018). This extends the philosophy from its limitations of retrospective studies of existing artifacts to a generative outlook of investigating new or speculative design artifacts. Investigating with artifacts like the counterfactual artifacts of the Morse Things or Tilting Bowl delivers on the promise of postphenomenology to understand things and technologies free of "pre-given normative frameworks" that focus on preconceived behaviors and norms. These normative frameworks obscure less visible or alternative understandings of how mediation occurs with artifacts.

In addition to design artifacts, HCI brings in-depth and innovative empirical methodologies that can be finely tuned to studying the relations between humans and technology. This, in turn, augments existing postphenomenology methods for studying technologies. As a result, HCI and design research can deeply engage the matter of technological mediation empirically. This approach can surface concrete, particular, and detailed accounts of human-technology relations that hold implications that can either richly affirm or problematize postphenomenology concepts.

Acknowledgments

We thank our expert participants for their generosity and effort and those from the Everyday Design Studio and Material Matters who helped with this project. Funded by NSERC and SSHRC.

Notes

1. This chapter is largely based on the study discussed in Wakkary et al. (2017) and extends the ideas of "displacement" and "withdrawal" discussed in Wakkary, Hauser, and Oogjes (2018).
2. Defamiliarization is originally a literary theory device (see Shklovskij 1991) to have readers examine their assumed interpretations of known and familiar experiences. The literary critic Frederic Jameson cited in Bleeker (2009) succinctly characterizes the aim to "defamiliarize and restructure our experience of our own present" (Jameson 1982). Bleecker is one account of utilizing defamiliarization in design fictions, and others have argued for it as a critical inquiry approach within HCI and domestic contexts in particular (see Bell and Dourish 2007).
3. We completed a first study of philosophers living with the Tilting Bowl, which was a three-month study (see Wakkary et al. 2018), and we are currently engaged in a second study, which has a duration of twelve months.

REFERENCES

Baird, Davis. 2004. *Thing Knowledge: A Philosophy of Scientific Instruments*. Berkeley, Calif: University of California Press.

Bell, Genevieve and Paul Dourish. 2007. "Yesterday's Tomorrows: Notes on Ubiquitous Computing's Dominant Vision." *Personal Ubiquitous Computing* 11 (2): 133–143. https://doi.org/10.1007/s00779-006-0071-x.

Bennett, Jane. 2009. *Vibrant Matter: A Political Ecology of Things*. Durham North Carolina: Duke University Press.

Bleeker, Julian. 2009. "Design Fiction: A Short Essay on Design, Science, Fact and Fiction." http://drbfw5wfjlxon.cloudfront.net/writing/DesignFiction_WebEdition.pdf.

Bogost, Ian. 2012. *Alien Phenomenology, or What It's Like to Be a Thing*. Minneapolis, MN: University of Minnesota Press.

Borgmann, Albert. 1987. *Technology and the Character of Contemporary Life: A Philosophical Inquiry*. Chicago: University of Chicago Press.

Comber, Rob, Anja Thieme, Ashur Rafiev, Nick Taylor, Nicole Krämer, and Patrick Olivier. 2013. "BinCam: Designing for Engagement with Facebook for Behavior Change." In *Human-Computer Interaction—INTERACT 2013: 14th IFIP TC 13 International Conference, Cape Town, South Africa, September 2–6, 2013, Proceedings, Part II*, edited by Paula Kotzé, Gary Marsden, Gitte Lindgaard, Janet Wesson, and Marco Winckler, 99–115. Berlin, Heidelberg: Springer Berlin Heidelberg. https://doi.org/10.1007/978-3-642-40480-1_7.

Crabtree, Andy and Peter Tolmie. 2016. "A Day in the Life of Things in the Home." In *Proceedings of the 19th ACM Conference on Computer-Supported Cooperative Work & Social Computing*, 1738–1750. CSCW '16. New York: ACM Press. https://doi.org/10.1145/2818048.2819954.

Davoli, Lorenzo and Johan Redström. 2014. "Materializing Infrastructures for Participatory Hacking." In *Proceedings of the 2014 Conference on Designing Interactive Systems*, 121–130. DIS '14. New York: ACM Press. https://doi.org/10.1145/2598510.2602961.

Fallman, Daniel. 2011. "The New Good: Exploring the Potential of Philosophy of Technology to Contribute to Human-Computer Interaction." In *Proceedings of the SIGCHI Conference on Human Factors in Computing Systems*, 1051–1060. CHI '11. New York: ACM Press. https://doi.org/10.1145/1978942.1979099.

Ganglbauer, Eva, Geraldine Fitzpatrick, and Rob Comber. 2013. "Negotiating Food Waste: Using a Practice Lens to Inform Design." *ACM Transactions on Computer-Human Interaction* 20 (2): 1–25. https://doi.org/10.1145/2463579.2463582.

Gaver, William W., Jacob Beaver, and Steve Benford. 2003. "Ambiguity as a Resource for Design." In *Proceedings of the SIGCHI Conference on Human Factors in Computing Systems*, 233–240. CHI '03. New York: ACM Press. https://doi.org/10.1145/642611.642653.

Giaccardi, Elisa, Nazli Cila, Chris Speed, and Melissa Caldwell. 2016. "Thing Ethnography: Doing Design Research with Non-Humans." In *Proceedings of the 2016 ACM Conference on Designing Interactive Systems*, 377–387. DIS '16. New York: ACM Press. https://doi.org/10.1145/2901790.2901905.

Harman, Graham. 2010. *Towards Speculative Realism: Essays and Lectures*. Winchester: Zero Books.

Hauser, Sabrina, Doenja Oogjes, Ron Wakkary, Peter-Paul Verbeek, Audrey Desjardins, Henry Lin, Matthew Dalton, Markus Schilling, and Gijs De Boer. 2018. "An Annotated Portfolio on Doing Postphenomenology Through Research Products." In *Proceedings of the 2018 Conference on Designing Interactive Systems*, 459–471. DIS '18. New York: ACM. doi: https://doi.org/10.1145/3196709.3196745.

Hauser, Sabrina, Ron Wakkary, William Odom, Peter-Paul Verbeek, Audrey Desjardins, Henry Lin, Matthew Dalton, Markus Schilling, and Gijs De Boer. 2018. "Deployments of the Table-Non-Table: A Reflection on the Relation Between Theory and Things in the Practice of Design Research." In *Proceedings of the SIGCHI Conference on Human*

Factors in Computing Systems. CHI '18. New York: ACM. Paper 201, 13 pages. DOI: https://doi.org/10.1145/3173574.3173775.
Heidegger, Martin. 1962. *Being and Time.* New York: Harper.
Ihde, Don. 1990. *Technology and the Lifeworld: From Garden to Earth.* Bloomington: Indiana University Press.
Ihde, Don. 1993. *Philosophy of Technology: An Introduction.* Paragon Issues in Philosophy. New York, NY, USA: Paragon House Publishers.
Jameson, Fredric. 1982. "Progress Versus Utopia; or, Can We Imagine the Future? (Progrès Contre Utopie, Ou: Pouvons-Nous Imaginer l'avenir)." *Science Fiction Studies* 9 (2): 147–158.
Keeney, Chris. 2014. *PetCam: The World Through the Lens of Our Four-Legged Friends.* New York: Princeton Architectural Press.
Latour, Bruno. 1987. "Where Are the Missing Masses? The Sociology of a Few Mundane Artifacts." In *The Social Construction of Technological Systems: New Directions in the Sociology and History of Technology*, 225–258. MIT Press.
Latour, Bruno. 1992. "Where Are the Missing Masses? The Sociology of a Few Mundane Artifacts." In *Shaping Technology/Building Society: Studies in Sociotechnical Change*, edited by Wiebe E. Bijker and John Law, 225–258. Inside Technology. Cambridge, Mass: MIT Press.
"Mytoaster (@mytoaster) | Twitter." n.d. https://twitter.com/mytoaster?lang=en (Retrieved April 6, 2017).
Odom, William and Ron Wakkary. 2015. "Intersecting with Unaware Objects." In *Proceedings of the 2015 ACM SIGCHI Conference on Creativity and Cognition*, 33–42. C&C '15. New York: ACM Press. https://doi.org/10.1145/2757226.2757240.
Odom, William, James Pierce, Erik Stolterman, and Eli Blevis. 2009. "Understanding Why We Preserve Some Things and Discard Others in the Context of Interaction Design." In *Proceedings of the SIGCHI Conference on Human Factors in Computing Systems*, 1053–1062. CHI '09. New York: ACM Press. https://doi.org/10.1145/1518701.1518862.
Odom, William, Ron Wakkary, Youn-kyung Lim, Audrey Desjardins, Bart Hengeveld, and Richard Banks. 2016. "From Research Prototype to Research Product." In *Proceedings of the 2016 CHI Conference on Human Factors in Computing Systems*, 2549–2561. CHI '16. New York: ACM Press. https://doi.org/10.1145/2858036.2858447.
Oogjes, Doenja and Ron Wakkary. 2017. "Videos of Things: Speculating on, Anticipating and Synthesizing Technological Mediations." In *Proceedings of the 35th Annual ACM Conference on Human Factors in Computing Systems*. CHI'17. Denver, CO: ACM Press.
Pierce, James. 2009. "Material Awareness: Promoting Reflection on Everyday Materiality." In *CHI '09 Extended Abstracts on Human Factors in Computing Systems*, 4459–4464. CHI EA '09. New York: ACM Press. https://doi.org/10.1145/1520340.1520683.
Pierce, James and Eric Paulos. 2011. "A Phenomenology of Human-Electricity Relations." In *Proceedings of the SIGCHI Conference on Human Factors in Computing Systems*, 2405–2408. CHI '11. New York: ACM Press. https://doi.org/10.1145/1978942.1979293.
Pierce, James and Eric Paulos. 2013. "Electric Materialities and Interactive Technology." In *Proceedings of the SIGCHI Conference on Human Factors in Computing Systems*, 119–128. CHI '13. New York: ACM Press. https://doi.org/10.1145/2470654.2470672.
Pschetz, Larissa and Richard Banks. 2013. "Long Living Chair," 2983. ACM Press. https://doi.org/10.1145/2468356.2479590.
Rosenberger, Robert and Peter-Paul Verbeek. 2015. "A Field Guide to Postphenomenology." In *Postphenomenological Investigations: Essays on Human-Technology Relations*, edited by Robert Rosenberger and Peter-Paul Verbeek, 9–41. Lanham: Lexington Books.

Shklovskij, Viktor Borisovic. 1991. *Theory of Prose*. 1. ed., 2., corr. Pr. Elmwood Park, IL: Dalkey Archive Press. https://www.dalkeyarchive.com/product/theory-of-prose/.

Tromp, Nynke, Paul Hekkert, and Peter-Paul Verbeek. 2011. "Design for Socially Responsible Behavior: A Classification of Influence Based on Intended User Experience." *Design Issues* 27 (3): 3–19. https://doi.org/10.1162/DESI_a_00087.

Verbeek, Peter-Paul. 2005. *What Things Do: Philosophical Reflections on Technology, Agency, and Design*. University Park, PA: Pennsylvania State University Press.

Verbeek, Peter-Paul. 2008. "Cyborg Intentionality: Rethinking the Phenomenology of Human–Technology Relations." *Phenomenology and the Cognitive Sciences* 7 (3): 387–395. https://doi.org/10.1007/s11097-008-9099-x.

Verbeek, Peter-Paul. 2009. *Moralizing Technology: On the Morality of Technological Artifacts and Their Design* in Readings in the Philosophy of Technology. Lanham: Rowman & Littlefield Publishers, 226–242.

Verbeek, Peter-Paul. 2011. *Moralizing Technology: Understanding and Designing the Morality of Things*. Chicago; London: University of Chicago Press.

Wakkary, Ron, Audrey Desjardins, and Sabrina Hauser. 2015. "Unselfconscious Interaction: A Conceptual Construct." *Interacting with Computers* 28 (4): 501–520. https://doi.org/10.1093/iwc/iwv018.

Wakkary, Ron, Sabrina Hauser, and Doenja Oogjes. 2018. "Displacement: Attending to the Role of Things in Theories of Practice Through Design Research." In *Social Practices and More-than-Humans*, 151–171. London: Palgrave Macmillan.

Wakkary, Ron, Doenja Oogjes, Henry Lin, and Sabrina Hauser. 2018. "Philosophers Living with the Tilting Bowl." In *Proceedings of the SIGCHI Conference on Human Factors in Computing Systems*. CHI '18. New York: ACM, Paper 94, 12 pages. doi: https://doi.org/10.1145/3173574.3173668.

Wakkary, Ron, William Odom, Sabrina Hauser, Garnet Hertz, and Henry Lin. 2015a. "Material Speculation: Actual Artifacts for Critical Inquiry." In *Proceedings of the Fifth Decennial Aarhus Conference on Critical Alternatives*, 97–108. AA '15. Aarhus: Aarhus University Press. http://dx.doi.org/10.7146/aahcc.v1i1.21299.

Wakkary, Ron, William Odom, Sabrina Hauser, Garnet Hertz, and Henry Lin. 2015b. "Material Speculation: Actual Artifacts for Critical Inquiry." *Aarhus Series on Human Centered Computing* 1 (1): 12. https://doi.org/10.7146/aahcc.v1i1.21299.

Wakkary, Ron, Doenja Oogjes, Sabrina Hauser, Henry Lin, Cheng Cao, Leo Ma, and Tijs Duel. 2017. "Morse Things: A Design Inquiry into the Gap Between Things and Us." In *Proceedings of the 2017 Conference on Designing Interactive Systems*, 503–514. DIS '17. New York: ACM Press. https://doi.org/10.1145/3064663.3064734.

12 REVEALING RELATIONS OF FLUID ASSEMBLAGES

Heather Wiltse

Contemporary things can often be characterized by their digital, networked, computational, "smart" character. They are more like *fluid assemblages* (Redström and Wiltse 2015; Wiltse, Stolterman, and Redström 2015; Redström and Wiltse 2019) than ordinary physical things. They are *fluid* in that their forms and functions change over time and across contexts and users, and *assemblages* in that they are emergent entities composed of a variety of interconnected and heterogeneous components that can still retain their identity when combined.[1] These include various web services and APIs, platforms, infrastructures, and of course both computational and physical components. Yet in use they tend to present themselves as stable, coherent things. It is as (at least provisionally) stable things that they enter into experience, as well as more philosophical analysis.

Human-thing relations are typically understood on the basis of intentionality—a thing never existing on its own but rather always as a thing for a human in a world. This is also the basic framing of perspectives in design that are focused on user experience, which have also been strongly influenced by phenomenology. Yet contemporary computational technologies fundamentally challenge this basic epistemological grounding in that much of what they are and do is not present to experience through use, and they can also involve other types of actors and intentional relations. These issues present fundamental challenges for understanding—both practically and philosophically—what they are and what they do, how they relate to us and to each other.

Within a (post)phenomenological framework, the basic intentionality relation is one in which a human subject is directed toward some aspect of the world that provides the content of that experience. Often this is toward or through a mediating technology, such that the basic unit of analysis becomes *I-technology-world* relations in postphenomenology (Ihde 1990). This type of analysis can be very productive when it comes to understanding human experience in relation to technologies; yet, it also leaves out significant aspects of what contemporary networked computational technologies are and what they actually do in the world of human affairs. Understanding these other aspects requires addressing other types of relations and multiple intentionalities—here called *multi-intentionality*.[2] Here, we need also the sensitivities of alien phenomenology (Bogost 2012) as method, investigating and speculating about the view from things.

The purpose of this chapter is to identify different forms of intentional relations that are at play in contemporary digital networked technologies, and associated implications and consequences for the roles that these designed things play in the world. These include relations among component parts and services; relations of corporate surveillance mediated through acts of use; personalized adaptation of things to humans (or at least humans rendered as data objects); and relations (albeit of a somewhat different kind) between what things do and how they appear. It is an exercise in revealing relations, and the ways in which they can reveal us.

Intentional Relations

Four types of intentional relations will be explored here: multi-intentionality, reverse intentionality, multi-instability, and interfaces that conceal.[3]

Multi-intentionality within and across Assemblages

Things that are fluid assemblages are actually constituted by multiple components that have their own agencies and connect and relate to each other in a variety of ways that are typically not at all present to those "using" what appear to be coherent things. This happens both within specific things and across things and the larger infrastructures, platforms, and assemblages of which they are part.

For example, a variety of components, services, and associated trackers are involved in most web pages, some even essential for their functioning. These are often used to customize web content and services in various ways. This can be revealed by, for example, using the Ghostery browser extension[4] that shows the trackers that are active on a given web page. We can also see a similar dynamic at work in the Spotify[5] music player, in which featured content is customized based on account, location, time of day, season, and so on. Some of the content also comes from other sources, as in the case of artist information that is attributed to Rovi. Rovi is another worldwide service, now part of TiVo,[6] that provides quite sophisticated API services and data for media apps and websites, including those for metadata, remote access, search, and personalized recommendations. Digging further into the multiple layers of Spotify, we can arrive at a list of the various types of cookies and other technologies that work together and are used to configure how Spotify appears as a thing in relation to specific users. We can also consider the possibilities that users have to affect this relation. It is particularly noteworthy that Spotify states in their privacy policy that they do not respond to "Do Not Track" signals set by users; this constrains the possibilities that people have to effectively control how Spotify relates to them. In the same policy they also identify several different websites, some geographically specific, where users can indicate they wish to opt out of behavioral advertising online.

The Spotify privacy policy also indicates that they "work with advertising partners to serve advertisements on the Spotify Service," including Google Analytics by Google. They state:

> We may use vendors, including Google, who use first-party cookies (such as the Google Analytics cookie) and third-party cookies (such as the DoubleClick cookie) together to inform, optimise, and serve ads based on your past visits to our websites, including Google Analytics for Display Advertising. Google provides tools to manage the collection and use of certain information by Google Analytics at tools.google.com/dlpage/gaoptout and by Google Analytics for Display Advertising or the Google Display Network by using Google's Ads Settings at google.com/settings/ads.[7]

Spotify shows up as a music player when approaching it for use, but this is actually the manifestation of a much broader assemblage that includes other technologies, companies, websites, services, and interests.

As another more simple example, consider the case of waking up to find that your smartphone battery has depleted substantially overnight without your having used it. When this happens, one might suspect the cause to be something to do with various updates, syncing, notifications, and so on. Yet at the same time there is something almost uncanny about wondering something like "What was my phone doing overnight that made it lose 18 percent of battery charge?"—and not knowing exactly or being readily able to find out the answer.[8]

Things have lives of their own, to which we are only partially privy as users. This starts to require object-oriented ontology and alien phenomenology (Bogost 2012) as analytic method in order to account for the character and relations of things. Things have their own relations to each other and (like us) only ever access parts of them.

Reverse Intentionality

There is also often what might be called a "reverse intentionality" at play in which one's use of a thing becomes the object of perception for someone else. This might be considered a special case of multi-intentionality and one that has significant implications for the roles that things play in our lives (and for whose benefit).

For example, read receipts or "seen by" badges make the fact that a message has been received and displayed (and indeed probably read) by someone else visible. Or in another more unexpected case, I once received a letter from my bank with the opening line: "We've noticed recently that you haven't been opening emails we're sending you." The letter goes on to instruct me to validate my email address and states that if I do not, they will stop sending me emails related to my account.

The basis of the bank's "noticing" and concern regarding my email-opening activity was unfounded, since I had in fact been receiving emails. However, I had also set my email preferences to not automatically display embedded HTML. Because the HTML was not loaded, there was no "read receipt" sent to the bank and they then assumed that I was not receiving email. Most commercial email senders do not complain explicitly when one does not display the content embedded in their emails, but this more striking example is also a reminder that these practices of effectively embedding read receipts to tally "impressions" have become standard practice for measuring the reach of marketing campaigns in industry. There are even now popular consumer email clients with this function, in which a tiny pixel, link, or something similar is embedded in sent emails. When these are downloaded by the recipient, this is registered by the client along with information about the location and device used (Merchant 2017).

This "noticing" of user behavior in web-connected applications and devices is becoming standard practice and also leading to new business models. For example, Netflix is able to monitor in detail what users stream, when, and on what devices; what they search for; where they pause, fast forward, and rewind; and so on. This capability allows them to deliver precisely personalized recommendations for what to watch. It is also what drove their decision to produce the original series "House of Cards," on the basis of knowing not only that they had a market for it but also that they would be able to deliver it to specific users in a way that was precisely tailored to the aspects of the show that would be most appealing to them (Carr 2013; Finn 2017).

In platforms such as Google and Facebook, "use" of the service is the means for profiling and serving targeted advertising. Facebook, for example, filters social interactions through its "profit-extracting sieve" (Plantin et al. 2017, 12) in order to generate a profit from packaging and selling users' attention. Google does something similar for search behavior. These services that appear as "things" available for use are thus quite different kinds of things to those on different sides of them. This can be seen clearly when comparing pages that face regular users versus those that face advertisers. Facebook, for example, is presented for users as a thing to "connect with friends and the world around you" (https://www.facebook.com/); for businesses, it is a way to "make meaningful connections with people to grow your business" (https://www.facebook.com/business). And this is not only on Facebook itself but also on the company's other apps (Instagram and Messenger) and Audience Network service, which delivers ads across sites and devices outside of Facebook[9]. Interestingly, there do not seem to be literal web links between these two sites, which represent two sides of what Facebook is as a thing.

Things can serve as mediators for the interests and activities of multiple actors. In this sense, reverse intentionality is often quite different from intentionality involved in use. Further, the presence of reverse intentionality is often concealed to varying degrees by user-facing surfaces. Things are designed to be engaging and easy to use for purposes that serve both users themselves and other actors, for whom people's ordinary use of a thing is a way to access them and extract other, very different kinds of value. This is the operational structure of surveillance capitalism, in which behavioral data is produced and reality more generally is mined for purposes of prediction and control in service of those able to pay in behavioral futures markets (Zuboff 2019). The humans formerly known as users have, according to common wisdom, already been turned into products to be sold; and now they have become resources to be stripped and used. Coming to grips with this situation requires developing sensitivity to reverse intentionality and the multiple relations and forms of use at play.

Multi-instability and Sources of Intentional Variations

One of the key concepts in postphenomenology is *multistability*, which refers to the possibility for humans to come into multiple relations to things. In other words, an object becomes stable as a particular kind of thing as it is understood and taken up into use by different people. Ihde (1990) develops this idea with the example of illusion drawings, which can be seen as different things, depending on how the gaze is focused. Multistability is a quite important concept because of the ways in which it makes space for human agency and intention in relation to things, thus countering more technologically deterministic narratives that tend to foreclose such possibilities. It also reflects a key methodological orientation of phenomenology proper, in which introduction of variations in experience is a way to get closer to understanding the "essence" of what a thing is.

However, when it comes to things that are fluid assemblages, there is also another set of dynamics in terms of how a stable relation can emerge from a variety of other possibilities and solidify in use. Specifically, in addition to humans being able to introduce variations in use, things that are fluid assemblages also introduce variations in how they present themselves. They can automatically customize themselves to specific users in specific contexts, showing up in slightly different configurations depending on the user and situation. "Smart" devices, for example, are customized from the first moments of setup, and the personalization only continues during use and through connection to other services with accounts and user data and machine learning capabilities. There are thus multiple sources and kinds of *in*stability—or what can be called *multi-instability.*

One simple way to notice this is to compare the smartphone home screens of different people, which might have quite different sets of apps organized in different ways and with different display settings (wallpaper and so on). This is a case of individuals not only achieving different relations to a smartphone as a thing but also literally relating to things that are formally and functionally different, even as the external shell might be the same. And of course other examples quickly become more complex.

Even more traditional kinds of things now come with software that is continuously updated by the manufacturer. For example, Sonos claimed that their speakers would "keep getting better" because of software updates that "mean the product you buy today will be even better tomorrow."[10] Digital networked things can also add connections to other applications and devices that change what they are able to do and the ways and places in which one's use of the thing is registered (as discussed previously under the heading of "reverse intentionality").

Another widespread mechanism of multi-instability is that of A/B testing, a now-standard method of developing digital products that involves testing the performance of different configurations ("A" and "B") against certain performance metrics at an often massive scale. While this might be seen as temporary testing in order to arrive at the final standard form, it now occurs with such scale and regularity in ongoing development that it is difficult to identify any particular instantiation as being the truly "final" one—it is rather the state of flux that is the norm. One of the major providers of A/B testing services, Optimizely, claimed that they help clients to "transform their customers' experience."[11] This is marketing hype gone literal, actually transforming the personalized content customers experience.

Introduction of variations is the engine that drives increasing understanding in phenomenology, but this is not really possible in the case of fluid assemblages. Everyone literally sees different things customized for them, and only a small part of a thing is revealed during use. Rather than increased revelation through use, we find increased—and increasingly precise—targeting and contextual customization. The multi-instability of things thus limits individual understanding as well as intersubjective understanding among people who have relations with presumably the "same" object or kind of object.

Interfaces That Conceal

Many aspects of what things are and do are—by design—concealed by user-facing interfaces. A key purpose of an interface is to present attractive and easy-to-use surfaces while concealing underlying complexity. For example, consider noted design theorist Klaus Krippendorff's comments:

> Interfaces constitute an entirely new kind of artifact, a human-technological symbiosis that cannot be attended to without reference to both. For designers, a key concern is that interfaces are *understandable*. Users' understanding need not be "correct" as intended by the producer, engineer, or designer of the technology. It needs to go only as far as needed for users to be able to interact with that technology as *naturally* and effortless as possible, without causing disruptions and reasons to fear failure. (Krippendorff 2006, 8)

And as design theorists Tony Fry and Clive Dilnot state in their "Manifesto for Redirective Design" (2003, 99):

The history of design is in many ways a history of concealing in the act of revealing—the increasing prominence of facade design in commercial architecture, the expansion of packaging design for graphics, the industrial design profession's activity of wrapping products in style, the expressive power of fashion as it exposes or hides the body—these are just a few example [*sic*] of this. What is concealed from view is not just what underlies appearances but equally the meaning of ways of knowing and acting.

It is also during use that the character of things is withdrawn in a more fundamental sense, as they take up their role as "equipment." Heidegger's ([1953] 2010) famous observations about how a tool withdraws from awareness during skilled use have become a common design goal. In other words, the idea is that people should be able to focus on what it is they want to do rather than on the tool used to do it. This goal is often explicit, as in Apple's advertising for the iPhone X, which asserts that it is "so immersive the device itself disappears into the experience."[12]

Now, it is probably safe to say that the experiences that Apple has in mind here are ones of communicating, creating, reading, downloading, listening, sharing, watching, playing, and similar types of activities that feature in their slick presentations to consumers. But there is much more at stake in how people use devices and the applications on them. For example, returning to the case of Spotify (which is an app that can be installed on an iPhone), the activity of listening to music is the mechanism by which Spotify very precisely packages and sells the attention of users to advertisers. The scale and precision involved are staggering: Spotify highlights that they have a "100% logged-in audience" and collect billions of data points every day. Further, they state:

> This user engagement fuels our streaming intelligence—insights that reflect the real people behind the devices. These real-time, personal insights go beyond demographics and device IDs alone to reveal our audience's moods, mindsets, tastes and behaviours.[13]

While people who use the free ad-supported version of Spotify are certainly aware that they are on the receiving end of advertising, the fact that Spotify is supposedly revealing their "moods, mindsets, tastes and behaviours" is less obvious during use. In fact, there are what might be seen as two complementary "sides" at play in Spotify, and other fluid assemblages more generally. On the one hand, personalized recommendations and adaptive configurations offer benefits during use that can be experienced as quite positive, and indeed as what we might now come to expect from increasingly sophisticated things; on the other hand, it is this very personalization that enables increasingly precise data collection and "insights" that can be extrapolated from it. Use of these kinds of things provides a certain kind of value to those who use them (e.g., music in the case of Spotify), while this use provides another kind of value to others in the larger sociotechnical systems in which they operate. This dual character of things that are fluid assemblages is concealed by interfaces that package them as engaging and even attentive things available for use.

Practically, digital things have always been concealed, or at least not completely revealed, by user-facing interfaces; more fundamentally, they withdraw during use in their role as equipment. The concealment by interfaces in both of these senses provides significant barriers to understanding what things really are and do.

On the Need for Breakdown

In the face of interfaces that conceal what things really do and styles of use that encourage perhaps too little conscious thought, how might it be possible to come to a place that allows better understanding and reflection? One concept that seems theoretically and practically promising is that of *breakdown*. There is a need to analytically break down things into the component parts of the assemblage in order to find out what is actually going on, as well as to break down the easy utility we may find in relation to them during use in order to achieve a more critical vantage point. It might also be necessary to break down ordinary ideas of the things we use that are based on their user-facing shells, in order to instead build up a more complex understanding of things that include their different kinds of components, users, and types of use. Breakdown is the inverse of Heideggerian withdrawal during use, the moment when a thing instead comes to presence as a broken, obstinate, misbehaving tool. If withdrawal means things retreating into the shadows, moments of breakdown trigger a metaphorical floodlight.

However, in looking for opportunities for constructive breakdown, we arrive at a more fundamental problem. Fluid assemblages cannot actually properly be revealed during breakdown, as in how a hammer can come to presence as a (malfunctioning) object when it ceases to be equipment taken up into use. The idea that a computer can reveal itself properly as an object or a thing is not viable. Rather than revealing themselves and coming into objective presence during breakdown, fluid assemblages tend to just disappear. Think, for example, of a computational device that becomes "bricked," and in losing its functionality simply becomes nothing more than a hunk of relatively useless material.

Or, in moments of less complete breakdown, rather than disappearing altogether, fluid assemblages can *lose face*—specifically, the interface that is supposed to enable people to use something but not necessarily explain how it actually works. To use Goffman's (1959) dramaturgical metaphor for human interactions, we could think of these as cases where the performance of things on the user-facing "stage" goes awry and the backstage processes begin to peek through the curtain in ways that can be confusing, yet oddly revealing.

For example, crash reports can give some indication of what goes on behind the scenes (and what went wrong), although they can also be fairly cryptic (to say the least) for those without the expertise to decipher them. Or for another example, when the Facebook website is slow to load for some reason (perhaps a slow computer or network connection), it is sometimes possible to briefly see text in placeholder boxes for images just before they load of the format "Image may contain: …." Some examples that I have seen and been able to capture include "2 people, people smiling, people sitting and living room"; "1 person, sitting, child"; and "2 people, people sitting, child, tree, outdoor." Now we can reasonably assume that these tags might be there for accessibility purposes in order to make it possible for those with visual impairment to somehow experience the image content shared on the site. However, it is also interesting to reflect on how the content shared on Facebook is processed and speculate about the other uses to which this type of (presumably) automatically generated data might be put, either now or in the future.

The causes of breakdown can also be interesting. Whereas we as users of more traditional things typically have the dubious privilege of being one of the main causes of breakdown as we use or misuse things, when it comes to fluid assemblages, other actors can also enter the picture. For example, in one case the maker of an IoT garage door opener bricked a customer's device after he published a negative review, by denying his particular unit server connection (Gallagher 2017). It is now becoming common for relations between producers and consumers to be ongoing through accounts and services, but in this case, that relation became even more personal through the mediation of the connected garage door opener. Whereas typical use would ideally be quite effortless and seemingly local, this rather forceful breakdown reveals its character as distributed assemblage involving multiple components, actors, and interests.

While true of all technologies, especially when it comes to things that are fluid assemblages, there is a need to understand technological mediation in a multifaceted sense that goes well beyond tool use to recognizing the roles that things play in complex political, social, and economic contexts and processes—especially when these processes are directly mediated by these things in an ongoing manner as part of their normal functioning. Uber is a good example for many kinds of current critical discussions, but one key aspect in this context is how it leverages intentionally designed information asymmetries to extract a profit from the difference between the routes and "upfront" fare prices it shows passengers and the most efficient routes shown to the drivers that are used to calculate their pay (Kravets 2017). We can also approach Facebook as a thing that requires outsourced and low-paid workers to filter out images of beheadings and other extremely offensive and disturbing content, and in a work environment where the Facebook ideology of "open sharing with the world" strictly does not apply (the workers have in fact been forbidden from speaking about their highly distressing work with anyone, even each other) (Chen 2014; Newitz 2017; Newton 2019). Or, to take another example, the Google Assistant is similarly powered by outsourced, low-paid linguistic experts (Wong 2019). This "ghost work" (Gray and Suri 2019) is a vital component of many digital services, but also another one meant to be concealed.

While things that are fluid assemblages cannot come to presence during breakdown in quite the same way as more purely physical tools, they also offer other opportunities for gaining insight and different vantage points. These opportunities are closely related to their character as assemblages. The fact that they are made up of various components and connections provides an opportunity for at least analytically laying out some of their moving parts and identifying the flows of data that they both utilize and feed. On the other hand, the fact that they are also themselves components in larger assemblages (not to mention the fact that they enroll their "users" in these assemblages as well) means that another key analytic move is to identify connections to other larger sociotechnical (infra) structures, processes of value creation, and interests—and, indeed, intentions and intentionalities. Another more practical move is to be the cause of breakdown as a user/value-producing component part in a system, disrupting the utility of things as revealers of user activity for others' profit by employing tactics of obfuscation (Brunton and Nissenbaum 2015), opting out, or similar.

The Mediating Role of Things in Surveillance Capitalism

Just as there is a need to break down assembled things in order to better understand what goes on behind their user-facing surfaces, there is also a need to zoom out and contextualize them within the larger sociotechnical and socioeconomic systems of which they are part. In fact, one of their more distinctive features is just how connected they are to larger systems and processes. Whereas things have often been constitutive of larger interconnected systems—such as cars that require highway networks, gas stations spaced at regular intervals, and so on—things are now often literally connected through digital networks and associated data flows. For example, the Waze app[14] allows drivers to share real-time traffic and road condition information with other users (explicitly involving multiple intentionalities of a user orienting toward the app as a thing providing useful information, but also as a thing gathering data about the world through one's use of it). The utility of the app is based on the real-time aggregation of data—in other words, not the thing itself, but the system in which it is a component part. In this example this dynamic is explicit, but in others, less so as the benefits of the system are more disproportionately distributed.

In fact, it could even be said that models in which data collected are fed directly back in ways that are beneficial for the very same acts of use are the exception rather than the rule. Laying out the contours of the new data economy in a 2017 article, *The Economist* states:

> Data are to this century what oil was to the last one: a driver of growth and change. Flows of data have created new infrastructure, new businesses, new monopolies, new politics and—crucially—new economics. Digital information

is unlike any previous resource; it is extracted, refined, valued, bought and sold in different ways. It changes the rules for markets and it demands new approaches from regulators. Many a battle will be fought over who should own, and benefit from, data. (*The Economist* 2017a)

Data is collected as a valuable resource, often without knowing exactly to which uses it might be put in the future or what kinds of insights it might generate. It is also collected in real time, and by an exploding number of sources. All sorts of devices are becoming "connected," allowing for more and different kinds of functionality but also, crucially, becoming themselves sites at which data is produced. This is a key dynamic of *platform capitalism*, a term used to describe the centrality of data and platforms in the shift toward new economic models based on the underlying structure of surveillance mediated by digital technologies (Zuboff 2019). As Nick Srnicek describes the situation:

Data is the basic resource that drives these firms, and it is data that gives them their advantage over competitors. Platforms, in turn, are designed as a mechanism for extracting and using that data: by providing the infrastructure and intermediation between different groups, platforms place themselves in a position in which they can monitor and extract all the interactions between these groups. This positioning is the source of their economic and political power. (Srnicek 2017)

To this description we might add: while platforms serve as the "profit-extracting sieves" (Plantin et al. 2017) that channel interactions through them in ways that generate valuable data, things are the more proximal mediators of those interactions—the sites where phenomena in the world bump up against devices and their algorithmic logic that render them visible in other ways. This process is in no way natural or neutral, and it involves complex and layered sets of (designed) mediations that can make everyday activities visible (Wiltse 2014) and encode them in computational form (Alaimo and Kallinikos 2017).

At stake here are the ways in which we, both individually and collectively as societies, come to know about the world and what goes on in it, and the ways in which we ourselves and our activities are rendered visible (in the broadest sense) and machine-readable in data form. This is what Shoshana Zuboff (2019) refers to as the division of learning in society: who knows, who decides, and who decides who decides? *Dataveillance*—"the disciplinary and control practice of monitoring, aggregating, and sorting data" (Raley 2013, 124)—is the new norm and prevailing socioeconomic technique, while efforts to conceptualize digital labor (including the production of data) as such and to configure positions of meaningful choice and control regarding one's own data have lagged far behind (The Economist 2017a). These countervailing efforts also face what have become huge and entrenched actors and systems. The "choice" of whether or not to accept terms of service that involve collection and use of data arguably does not often meet what might be conceived as standards for being a meaningful one, since not accepting simply means not being able to use applications and services that have in many ways become effective infrastructure in collective social, economic, and political life (Plantin et al. 2017; Wittkower, present volume).

The "standard account" of designed things in an industrial context can be said to be one in which their primary functionality is in the utility they provide through use and the exchange value that stems from that. In the contemporary conditions of surveillance capitalism, this situation fundamentally changes. The utility of things in use becomes also the means of getting people to use things and thus enable the collection of data about that use. Data empires consolidate their positions in a "god's-eye view" through aggregating data from their multiple products and the connections between them (*The Economist* 2017b). This function of things is generally not present during acts of typical use of what appear to be discrete things. Although we might get glimpses of these dynamics when noticing ads that follow us between multiple sites and devices or apps that are remarkably insistent about getting us to go for a run, practice a language, share about our lives, check updates from others, or whatever else, connected things actively relate to us, but more than that, other entities and actors relate to us through these things. They are machines, to use Levi Bryant's (2014) terminology and ontology, that have their own agencies and relations to other machines that are not accessible through normal acts of use.

When digital networked things relate to us, it is not only things but also other actors and agencies relating to us through the mediation of the thing. These two types of relating are also intertwined: the customization of things for specific people and situations can be a means of probing, testing, and gathering more detailed data about use and users. For example, Spotify featured playlists that are presented to fit a particular mood, time of day, day of the week, season, or activity can be a welcome offering of exactly the kind of thing a person wants to listen to in a particular moment; and the act of choosing to listen to a certain curated playlist then also becomes data that is fed back as a signal regarding what one is doing, thinking, and feeling. As Spotify puts it in their brand-facing pitch:

> We've found that how people stream actually tells us a lot about who they are. Our data team has identified five key streaming habits that can help you understand your audience, and better inform your planning. The most exciting part? This new research is starting to reveal the streaming generation's offline behaviors through their streaming habits.[15]

In using Spotify to listen to music, one does not quite get the sense that it is a highly sophisticated system used to collect data intended to reveal one's "mood, mindset, taste, and behavior" in real time. It is rather the opposite—this data collection system is packaged as an engaging and responsive thing available for use.

The significance of things in everyday life and activities has traditionally been experienced and treated analytically as a matter of the uses to which they are put. But things have changed—and so have their multiple uses, users, and relations. If breakdown can be a tactic for understanding and intervention, it might then also serve as an invitation to assemble differently, with care for relations and their consequences.

Conclusion

When using a thing that is a fluid assemblage, we literally become part of its making and also enter ourselves into assemblages that we use and that in many instances also use us. In order to properly care for human experience, we need to go beyond what is present to humans through use and consider how experience is affected in other ways as well. A key part of this agenda is learning how to follow the multiple relations and intentionalities at play in assembled things: breaking things down in order to understand their component parts and processes and contextualizing them in the larger systems in which they—and those who use them—are component parts.

Acknowledgments

The work described in this chapter developed in large part through collaboration with Johan Redström over a number of years, which I gratefully acknowledge. The final version of the chapter was also helped substantially by close reading and thoughtful comments by Holly Robbins. The work was funded in part by the Marianne and Marcus Wallenberg Foundation (research project grant MMW 2017.0058).

Notes

1. This conception of assemblages builds on the work of Deleuze and Guattari (Deleuze and Guattari 1987), DeLanda (2016), and Bryant (2014), among others.
2. This concept has been developed in more depth in Redström and Wiltse (2019).
3. The concepts of multiinstability and multiintentionality have also been developed previously in Redström and Wiltse (2019).
4. https://www.ghostery.com/
5. https://www.spotify.com/
6. https://business.tivo.com/
7. https://www.spotify.com/us/legal/privacy-policy/#s13, accessed 9 February 2018.
8. Reporter Kashmir Hill's in-depth investigative experiment in cutting the "big five" (Amazon, Facebook, Google, Microsoft, and Apple) out of her life through using a custom-built VPN is illuminating on these issues, especially when it comes to these tech giants that have their services deeply embedded in many web services (Hill 2019).
9. https://www.facebook.com/business/products/audience-network, accessed 9 February 2018.

10   https://www.sonos.com/en-gb/shop/play3.html, accessed 8 June 2017 but no longer displaying this text as of the current writing.
11   http://optimizely.com/, accessed 8 June 2017 but no longer displaying this text as of the current writing.
12   https://www.apple.com/iphone-x/, accessed 22 January 2018.
13   https://spotifyforbrands.com/en-GB/audiences/, accessed 22 January 2018.
14   https://www.waze.com/
15   https://spotifyforbrands.com/en-US/audiences/, accessed 9 February 2018.

References

Alaimo, Cristina and Jannis Kallinikos. 2017. "Computing the Everyday: Social Media as Data Platforms." *The Information Society* 4 (33): 175–191. https://dx.doi.org/10.1080/01972243.2017.1318327.

Bogost, Ian. 2012. *Alien Phenomenology, or, What It's Like to Be a Thing*. Minneapolis; London: University of Minnesota Press.

Brunton, Finn and Helen Nissenbaum. 2015. *Obfuscation: A User's Guide for Privacy and Protest*. Cambridge, MA; London: MIT Press.

Bryant, Levi R. 2014. *Onto-Cartography: An Ontology of Machines and Media*. Edinburgh: Edinburgh University Press.

Carr, David. 2013. "For 'House of Cards,' Using Big Data to Guarantee Its Popularity," February. http://www.nytimes.com/2013/02/25/business/media/for-house-of-cards-using-big-data-to-guarantee-its-popularity.html.

Chen, Adrian. 2014. "The Laborers Who Keep Dick Pics and Beheadings Out of Your Facebook Feed." *Wired*. October 23. https://www.wired.com/2014/10/content-moderation/.

DeLanda, Manuel. 2016. *Assemblage Theory*. Edinburgh: Edinburgh University Press.

Deleuze, Gilles and Félix Guattari. 1987. *A Thousand Plateaus: Capitalism and Schizophrenia*. Translated by Brian Massumi. Minneapolis; London: University of Minnesota Press.

The Economist. 2017a. "Data Is Giving Rise to a New Economy." *The Economist*. May 6. https://www.economist.com/news/briefing/21721634-how-it-shaping-up-data-givingrise-new-economy.

The Economist. 2017b. "The World's Most Valuable Resource Is No Longer Oil, but Data," May. https://www.economist.com/news/leaders/21721656-data-economy-demandsnew-approach-antitrust-rules-worlds-most-valuable-resource.

Finn, Ed. 2017. *What Algorithms Want: Imagination in the Age of Computing*. Cambridge, MA; London: MIT Press.

Fry, Tony and Clive Dilnot. 2003. "Manifesto for Redirective Design: Hot Debate." *Design Philosophy Papers* 1 (2): 95–103. doi: 10.2752/144871303X13965299301795.

Gallagher, Sean. 2017. "IoT Garage Door Opener Maker Bricks Customer's Product After Bad Review." *Ars Technica*. April 4. https://arstechnica.com/information-technology/2017/04/iot-garage-door-opener-maker-bricks-customers-product-after-bad-review/#p3.

Goffman, Erving. 1959. *The Presentation of Self in Everyday Life*. New York: Anchor Books.

Gray, Mary L. and Siddharth Suri. 2019. *Ghost Work: How to Stop Silicon Valley from Building a New Global Underclass*. Boston and New York: Houghton Mifflin Harcourt.

Heidegger, Martin. [1953] 2010. *Being and Time*. Translated by Joan Stambaugh. Albany, NY: SUNY Press.

Hill, Kashmir. 2019. "Life Without the Tech Giants: Goodbye Big Five." *Gizmodo*, January 22, 2019. https://gizmodo.com/life-without-the-tech-giants–1830258056.

Ihde, Don. 1990. *Technology and the Lifeworld: From Garden to Earth*. Bloomington: Indiana University Press.

Kravets, David. 2017. "Uber Said to Use 'Sophisticated' Software to Defraud Drivers, Passengers." April 6. https://arstechnica.com/tech-policy/2017/04/uber-said-to-use-sophisticated-software-to-defraud-drivers-passengers/#p3.

Krippendorff, Klaus. 2006. *The Semantic Turn: A New Foundation for Design*. Boca Raton, FL: CRC/Taylor & Francis.

Merchant, Brian. 2017. "How Email Open Tracking Quietly Took over the Web." *Wired*. December 11. https://www.wired.com/story/how-email-open-tracking-quietly-took-over-the-web/.

Newitz, Annalee. 2017. "Will Facebook Actually Hire 3,000 Content Moderators, or Will They Outsource?." *Ars Technica*. May 4. https://arstechnica.com/tech-policy/2017/05/facebook-promises-to-hire-3000-people-to-moderate-content/#p3.

Newton, Casey. 2019. "The Trauma Floor: The Secret Lives of Facebook Moderators in America." *The Verge*, February 25, 2019. https://www.theverge.com/2019/2/25/18229714/cognizant-facebook-content-moderator-interviews-trauma-working-conditions-arizona.

Plantin, Jean-Christophe, Carl Lagoze, Paul N. Edwards, and Christian Sandvig. 2017. "Infrastructure Studies Meet Platform Studies in the Age of Google and Facebook." *New Media & Society* 18 (1): 293–310. doi: 10.1177/1461444816661553.

Raley, Rita. 2013. "Dataveillance and Counterveillance." In "*Raw Data" Is an Oxymoron*, edited by Lisa Gitelman, 121–145. Cambridge, MA; London: MIT Press.

Redström, Johan and Heather Wiltse. 2015. "Press Play: Acts of Defining (in) Fluid Assemblages." Proceedings of Nordes 2015: Design Ecologies. http://www.nordes.org/opj/index.php/n13/article/view/432/407.

Redström, Johan and Heather Wiltse. 2019. *Changing Things: The Future of Objects in a Digital World*. London: Bloomsbury.

Srnicek, Nick. 2017. "The Challenges of Platform Capitalism: Understanding the Logic of a New Business Modelx." *Juncture* 23 (4): 254–257. doi: 10.1111/newe.12023.

Wiltse, Heather. 2014. "Unpacking Digital Material Mediation." *Techné: Research in Philosophy and Technology* 18 (3). Philosophy Documentation Center: 154–182. http://dx.doi.org/10.5840/techne201411322.

Wiltse, Heather, Erik Stolterman, and Johan Redström. 2015. "Wicked Interactions: (on the Necessity of) Reframing the 'Computer' in Philosophy and Design." *Techné: Research in Philosophy and Technology* 19 (1). Philosophy Documentation Center: 26–49. http://dx.doi.org/10.5840/techne201531926.

Wong, Julia Carrie. 2019. "'A White-Collar Sweatshop': Google Assistant Contractors Allege Wage Theft." *The Guardian*, May 29, 2019. https://www.theguardian.com/technology/2019/may/28/a-white-collar-sweatshop-google-assistant-contractors-allege-wage-theft.

Zuboff, Shoshana. 2019. *The Age of Surveillance Capitalism: The Fight for a Human Future at the New Frontier of Power*. London: Profile Books.

13 DESIGNING NETWORKS THAT REVEAL THEMSELVES

Holly Robbins

The world today is populated by networked technologies that are imbued with the capability to take note of the things that we do and try to identify patterns in our behaviors and preferences to adjust their own behavior accordingly. Colloquially, these are referred to as "smart" technologies. They're "smart" because the people who use them don't have to crank levers, push buttons, or program things for them to function; it is as if they possess their own intelligence and agency to operate themselves. Yet, even without these forms of exertion, we are still essential to how they function. These technologies learn about us, communicate that information with others, contribute our information to a larger data set, and evaluate it, and then the function of those artifacts is tailored to our individual idiosyncrasies. With these networked technologies, we interact not just with the artifact that's in our hands but also with an entire network behind it. Yet it is only outcome of the technology's use that is most accessible and revealed to us, not the networks that contribute to its function.

Organizational theorist Wanda Orlikowski aptly demonstrates this dynamic with the example of a Google web search (2007). When I submit a search query, one could assume that Google scours the internet to find the best answer to my question. But the internet is massive, and what is not apparent is the work that Google does to deliver results that are tailored to that specific person submitting the question. The search engine's algorithms favor some content over others by considering my personal browser history, location, and various other indexing and ranking of pages and content that Google does internally. Google is a powerful technology that draws on a network of various entities and materials to deliver the result that Google presumes is best suited to me based on its assessment of who I am, as well as what may be Google's agenda.

In effect, I become a node in Google's network, contributing resources (such as personal data) to shaping the network itself. In addition to tailoring my own results, my data ultimately contributes to informing the results of another person making a Google search. The search results therefore are a reflection of how the networks behind Google's search engine relate to me. And in turn I relate to its results, which inform my understanding of the world. This can have profound effects, as is the case with "filter bubbles" (Eslami et al. 2015).

However, this dynamic of mutual relating is not made apparent. For example, the interfaces of these technologies don't reveal these various processes of relating occurring. In the previous chapter in this volume, Heather Wiltse provides a comprehensive account of a variety of different ways, or intentions, that technologies can relate to the people using them. She suggests that it is only when these technologies break down that we become aware of these intentions or forms of relating.

This is fundamentally a question of how these artifacts are designed. Choices have been made in the development and design of these technologies to hide these networks, their intentions, how they relate to people, and people's roles in these networks from the people who make use of them. As a result, the way that these technologies relate to us is not being revealed by their design.

The question that this chapter asks is: How can the design of these technologies reveal *these networks to make them more relatable to the people making use of, and are a part of, them?* Masking the complexity of these technologies is generally encouraged in a product design context to promote their accessibility and usability for the general public. Not everyone can program, assemble, or debug a laptop, but they have been designed in a way so that most of us can use one despite this lack of technical expertise. It is therefore *counterintuitive within a design context to seek to reveal* the complexity of these contemporary technologies. This chapter considers proactive approaches through which the design of these technologies can reveal, to the layperson making use of them, a bit of the complexity of these networked connections and how they relate to us.

To approach this objective, this chapter will first turn to philosophy of technology to frame this concept of "relatability." Specifically, Albert Borgmann's device paradigm is used to decipher what about complex technologies challenges people's ability to relate to them (Borgmann 1984). This framing will then be contextualized within interaction design practices, in an effort to understand how design can potentially contribute to making connected or networked technologies more relatable. Two conceptual designs will be described to illustrate how the design of connected or networked products can promote relatability between the person and the technology. This chapter will close with an observation that relatability is not only in revealing the complexity of the technology but also in putting users in an active role of making sense of the technology and their relationship with it.

Framing Relatability with the Device Paradigm

Albert Borgmann's work on the "device paradigm" provides a framing to approach this question of how people relate to technologies in terms that are very relevant to design. In particular, he argues that the breakdown in our ability to relate to technology occurs when the *ends* of the technology (the outcome of the technology's use) are separated from its *means* (the aspects of the technology responsible for the way it works) (Borgmann 1984). In doing so, people are becoming disconnected from the task the technology performs on a bodily, social, and environmental level. Borgmann suggests that this separation results in not understanding how the technology works, the role the technology plays in our lives, and can even potentially lead to its overuse. As a consequence, our only mode of relating to, and engagement with, the technology is in consuming its output, or its *ends*. This is becoming increasingly commonplace among contemporary technologies.

Borgmann uses the example of heating technologies to illustrate the device paradigm. We once heated homes with fireplaces. The fire deeply draws on, or engages with, our bodily senses (bodily skills developed and sensorial experiences related to chopping and handling the wood and keeping the fire going), sociality (different roles in maintaining the fire, being the focal point of a home), and materiality (the way that different woods cut and burn). We can see here how the technology of the fire is positioned in these ecologies of different modes of human engagement. There is a lot of social and ecological context behind how this technology works and the role it plays in our lives. This changes with distributed heating, where a slight turn of the wrist on a thermostat offers heating on demand, dispensing it directly to individual rooms. There is no bodily engagement with the thermostat or the furnace. Nor is there social engagement around this technology; the heat can be enjoyed in the privacy of our own rooms. The thermostat in effect separates the *ends* (the output of the heat) from the *means* (how the heat is procured).

Borgmann crafted the device paradigm in the 1980s, well before networked technologies were pervasive. Networked technologies have new capabilities that limit our bodily, social, and material engagement, further separating the *ends* from the *means*. For example, the Nest thermostat doesn't require that slight turn of the wrist: it learns the patterns and preferences of the members of the household, anticipates the demands for heating, and supplies it before being asked. The whole premise of the Nest is that it is a thermostat that does not require your engagement. It collects data from you, which is then contributed to a pool of other data; that pool of other data is evaluated, and then the behavior of the Nest is determined.

Like the earlier example of the Google search, the technology's *means* is configured in drawing data from the people who use it and analyzing that data within a larger data set. Then the technology's function (*ends*) is tailored specifically to that person making use of it. Because of the separation between the *ends* of its tailored services and the *means* of the data collection and interpretation enabled through its networked capabilities, there is a lack of context for the layperson of how the technology works and the role that that person plays in contributing to how it works. In this case relatability becomes lopsided. *The technology, through its networks and learning capabilities, is relating to us; but without much context to these activities, the users of these technologies are not provided the opportunity to relate to it.*

Framing Relatability within Interaction Design

Translating this framing of relatability from the device paradigm into the language of interaction design, Borgmann's critique is that the user's interactions with the technology are isolated and masked from the task the technology performs. A widely accepted tenet of interaction design practice is to strive to create "human-centric" technologies. Typically, this translates to designing the ways that we interact with technologies to prioritize human needs, such as being easy to use and efficient, and that strive to gratify the user (user satisfaction). With these priorities, design is used to offer smooth and seamless interactions and experiences that the person has with the technology. Generally speaking, this has been achieved with interfaces and encasements that mask or obfuscate the complexity of the technology to make it more palatable for the general user.

For example, the interaction of pushing a button to start a car, or even now merely having a key in one's pocket in proximity to the car, does little to reveal the complexity behind how an engine is turned on. On the other hand, the interaction of winding a crank to turn the engine of some of the earliest car models more closely references how that technology works and involves the person in joining the *ends* and the *means* together. However, winding a crank of a car to turn the engine isn't particularly easy, efficient, or gratifying when we're in a rush to get to work. Instead of having to engage with the mechanics of the technology to operate it, having a key that was already in your pocket turn the car on seems like an easier, efficient, and gratifying option, and therefore preferable design solution.

Predominate design conventions propagate the device paradigm; specifically, the distance between the interaction that we have with the technology (*means*) and the task that it performs (*ends*) that gives rise to the device paradigm. As Borgmann described, this deprives the user of that technology a sense of context of how it works and the effect of our involvement with it. As we turn to technologies that automate their function, this is exacerbated. With automation there are fewer opportunities for active interactions, and thus less context because the technology operates and runs itself. *These conventional design approaches aim to make the technology relatable by promoting usability of the technology. In obfuscating the complexities surrounding the technology, relatability is characterized by a lack of knowledge and context across a broader population.*

But, perhaps this approach to how we design technologies is not appropriate; it both exacerbates the device paradigm and is built on the premise that "relatability" is defined by collective ignorance, not necessarily understanding. This proposition echoes one made by Daniel Fallman, who questions the very conventions of "good design" that contribute to the device paradigm in the first place (Fallman 2009; Fallman 2011). He suggests that technologies should be developed not just with an eye toward their efficient use but that designers and developers should also work toward providing opportunities for the social implications of these technologies to be revealed, demonstrating the particular relationships and dependencies that they foster (Fallman 2011). Perhaps that which is conventionally understood to be "good design" in the first place, those that favor these efficient and gratifying interactions, isn't so good after all. This chapter suggests that there is an alternative to this institution of "good design," which isn't necessarily inconvenient or burdensome, as the "opposite" of these design conventions may suggest.

By framing the device paradigm as a consequence of design decisions, it becomes possible to explore how choices in the design of these networked technologies can circumvent the device paradigm. The following sections in this chapter explore possible approaches that can be taken with design to reveal the complexities of how these networked technologies relate to people, in a way that is accessible to the people using them. In doing so, relatability can become mutual between the technology and the person. The following sections of this chapter describe conceptual work in design research that offers perspective as to how to promote relatability with design (Koskinen et al. 2011).

Materializing Networks

If design materials are defined as the parts that can constitute an object, then the networks behind connected objects are a part of those materials (Robbins, Giaccardi, and Karana 2016; Robbins 2018). For those technologies with connectivity, networks are a part of the object and enable it to function and perform its task. Just as designers consider how to use wood or plastic and how people will interact with these materials when creating an object, so too should they consider networks and how people understand the ways they interact with them. Designers should consider networks as having a form that needs to be expressed through the interaction that people have with their product or service. Considering networks a material to be revealed is an attempt to contextualize the complexity of the technology. This aligns with Borgmann's framing of the device paradigm. The most promising intervention that design can offer will have to be in developing our understanding of how our interactions with networks as materials (*means*) are related to the way the technology functions and the outcome of its use (*ends*).

The material turn in interaction design seeks to find ways to understand the tangibility of intangible things, such as that which is often referred to as the "digital" (Wiberg and Robles 2010; Dourish and Mazmanian 2011; Rosner et al. 2012; Nansen et al. 2014). This has two purposes. First, this is to remind us that intangible things are rooted in material forms. Data is stored on servers and is connected with wires, which occupy physical space and require utilities such as electricity and air-conditioning to operate. This intangible thing of digital data has physical material properties in the form of server farms (Dourish 2014). Further, these material properties shape our interactions with them as the interactions that they can have with other objects. For example, if the air-conditioning fails on a server farm, a critical server will likely crash, making a website temporarily unavailable (Dourish and Mazmanian 2011).

A defining characteristic of networks is their relationality. The challenge is for design to represent that network and represent the user as a node within it. This then becomes a call to design researchers to take a step away from thinking about networks purely in human-centric or *ends*-centric terms, where the objective of the network is to deliver services to the human. Instead, such a perspective prompts us to design networks of artifacts that "consider how objects already exist in established networks of relationships with people and how this sociality can be incorporated in insulated, engaging, shared and meaningful ways" (Nansen et al. 2014).

A growing body of work within design research also encourages us to consider the perspective of the networks that objects exist within and the agency and social relationality that these objects possess (Nansen et al. 2014; Giaccardi et al. 2016; Wakkary et al. 2017; Giccardi and Wakkary et al. this volume). With objects as social agents, they have "specific properties, histories, affordances and relations," which impact humans as well as being defined by them (Nansen et al. 2014).

In the following sections, I turn two conceptual designs that provoke this question of how the design of a technology can reveal networks and our role within them. The first design approach does this by developing a product label with a symbolic language to this effect (Thingformation), while the second design approach conveys this positioning and relationality through dynamic interactions with the technology (The Transparent Charging Station).

Labeling Networks: Thingformation

Thingformation is a design concept created by the Belgian design agency beyond.io (beyond.io 2017). They were tasked to find a way to make the parties associated with an internet-connected product explicit (Afdeling Buitengewone Zaken et al. 2015). Thingformation is a product packaging label that communicates information that is not immediately apparent about the network behind a product (Figure 13.1). The label classifies a product on five distinct, although interrelated, qualities that refer to the networked complexity behind the object: type of encryption used with the product; the number of companies affiliated with the product; what body of laws regarding data protection the product is held accountable to; the expiration date of the product; and lastly, a graded evaluation of the trustworthiness of the company with regard to how they use their customer's data. This design concept aims to reveal the complexities of a connected object in the spirit of consumer protection. Thingformation is offered not only to help inform purchasing decisions but also to motivate industry standards and support the development of our sensibilities in navigating and understanding these complex technologies (Bihr 2017; Robbins and Just Things Foundation 2017).

It may be easy to overlook Thingformation's contribution to our greater objective of supporting relatability; it's merely a label. However, with this labeling system, Thingformation attempts to literally reveal the network, but in such a way as to promote its ability to be understood by a general audience. This is similar to

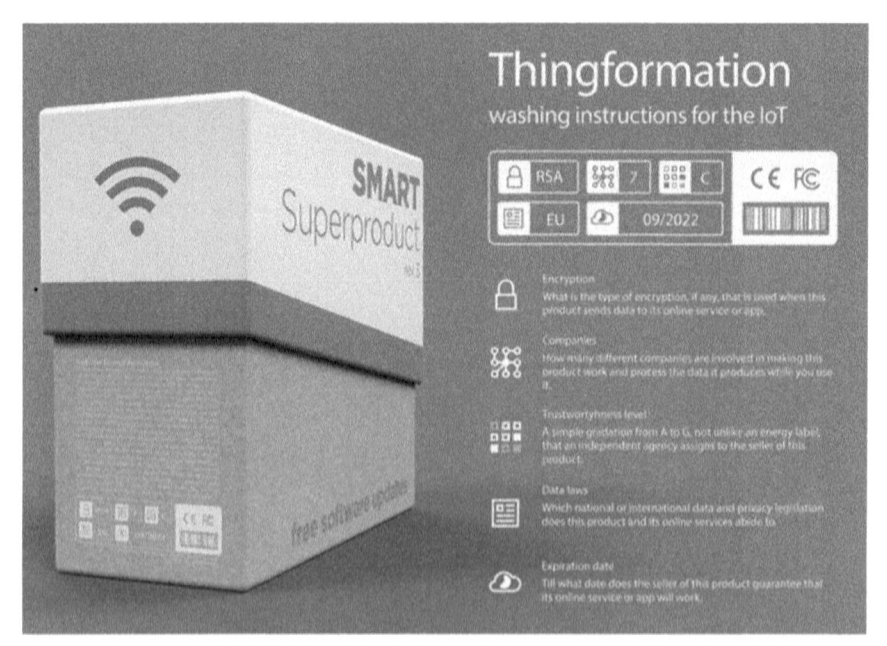

FIGURE 13.1 Thingformation, an IoT care labeling system for product packaging. Design and image: Beyond/IO.

how wash labels on clothing instruct people on how to care for the clothing with the complex washing machinery that they have in their homes.

Thingformation offers an avenue to reveal the network by creating an interface. This label is where people can "meet" the network that is being accessed through the object, which people are implicitly also becoming a node within. With this interface, Thingformation enables people not only to develop the awareness of the network they're now a part of but also to build upon this awareness to ideally develop an informed opinion of how they would like to inhabit this network.

Interacting with Networks: The Transparent Charging Station

The second conceptual design artifact that this chapter will discuss is the Transparent Charging Station. Commissioned by a Dutch energy company and designed by the Dutch design agency The Incredible Machine, this station attempts to reveal the network behind the electric energy grid of car charging station and make it interactive for its consumers (Amsterdam Institute for Advanced Metropolitan Solutions 2017; The Incredible Machine 2018).

In the not so distant future most cars will be electric, which will radically challenge existing practices surrounding how we fuel cars. Car batteries require more time to recharge than it takes to fill a car with petrol at the station. Additionally, under normal circumstances, conventional petrol stations always have a reservoir of fuel beneath the surface available on demand. These conventional petrol stations come to represent an example of the device paradigm. The station's design instead masks the complexity behind the infrastructure of refueling, leaving people with the luxury of knowing that the experience will be efficient, reliable, and easy. The availability of the petrol to refuel your car (*ends*) is offered without context to how it became available (*means*).

However, with an electric fueling infrastructure, this will change. The availability of electric energy at the "pump" will fluctuate in response to a number of factors: the demand on the electric grid at any particular moment, the availability of renewable resources, the weather, and what's already currently stored. There is a network of factors that will influence how and when the car can possibly be charged. To address this complexity in refueling electric cars on a mass scale, The Incredible Machine problematizes how to deliver this energy to cars, while conveying the network that is necessary to its functioning with the Transparent Charging Station.

Each charging port of the station has two dials (Figure 13.2, close up in FIgure 13.3, along the bottom). By turning these dials, drivers set the time that they require a charge and what percentage of their battery needs to be refilled. The station's interface responds to this by illustrating how the grid may or may not be able to satisfy this precise request. This interface resembles a Tetris matrix, which narrows and widens based on what resources are available, and is anticipated to be available, on the electric grid (Figure 13.3). The request being made of that particular station is accommodated into that matrix, fitting within the electric grid's overall constraints. The driver's request has to be balanced with those also being made by that particular station's other patrons. When turning the dials to select the percentage and time frame for that charge, drivers are making a negotiation with the constraints of what other people have requested as well as what is available on the network as a whole. The interface changes in a fluid and dynamic way in response to the requests being made and the constraints on it at that moment. In turning the dials, the driver can broker an arrangement between him and the network, experimenting with different plans and compromises.

In turning these dials, actively navigating, interacting, and negotiating with the algorithmic constraints, the user is directly relating to and contextualizing the network itself. As a user, you can see the impact your request has on the other nodes of the network, such as the others who are charging at that very pump, as the interface's Tetris-like screen changes its shape and color accordingly. Transparency is offered not to the extent where the algorithms are explicit. That would be too dense to be accessible to the layperson anyway, and therefore not transparent. Instead, through these dynamic interactions, people begin to form an understanding of the complexity of the network and their own role in it. This dynamically and interactively reveals the complexity of the network behind the charging station and offers people the context they need to navigate that complexity.

FIGURE 13.2 The Transparent Charging Station is an electric car charging station that allows people to negotiate how much of their battery is to be charged and within what time period according to the networked constraints of an electric energy grid. Design and image: The Incredible Machine.

With turning the dials in response to this representation of the availability of the grid, people are joining the ends and the means of the technology together, as Borgmann advocated for. People are now provided an active and engaging role in navigating the complex network behind the technology because design has made it legible for them.

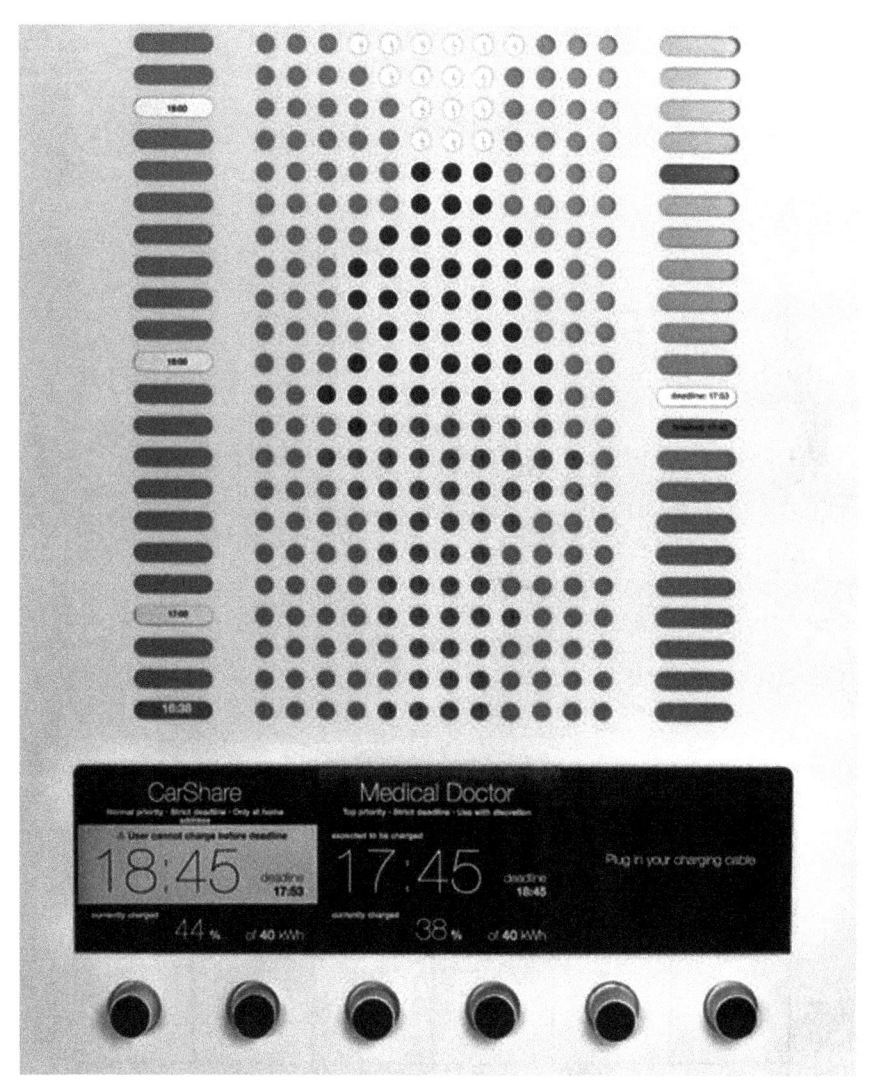

FIGURE 13.3 The interface of the Transparent Charging Station demonstrates the constraints and demands on the electric grid with a Tetris-like screen. The hourglass-like figure in the middle illustrates what energy is predicted to be available on the grid over the course of the day. This station has three different charging ports, each of which is controlled by two dials along the bottom. In turning the dials, a driver negotiates how much of her battery is to be changed and in what timeframe within the constraints of the energy predicted to be available on the grid. As the driver turns these dials, the screen also illustrates how their request impacts those made by the others charging at this port.

From Revealing to Relating

There is an important distinction that must be considered, which is that between revealing the complexity of the technology and making that complexity relatable to people. For the technology's complexity to be relatable, its design should support avenues that provide the opportunity for people to *make sense of* and contextualize the complexity. This complexity can not only be revealed to show that it exists but there must also be some contextualization. I argue that relatability can be supported with users taking an active role in navigating and making sense of that network.

Simply explicitly stating that the network exists isn't sufficient to making it relatable. This is the case with terms of service agreements. All the information regarding the object is available in the form of text to be read in extensive detail and often written in impenetrable legalese. If one manages to get through these agreements, which the majority of people don't, that information typically does not bear any relevance once we use the technology. How we convey the reality of the technology cannot be explicit or prescriptive. *Instead, design needs to offer modes of engaging with the complexity of the technology that is open-ended, supporting the general user's ability to develop a sense of context surrounding that complexity and also his or her ability to navigate that information.*

For example, the Transparent Charging Station doesn't provide its users with the code behind the algorithms that demonstrate the availability of the energy and how that can be satisfied within the constraints of the request being made. Instead, it shows the user what it does and is making it insightful. We see how the input from a request interacts with the system's constraints and how it impacts others. With that knowledge, the user can change his or her request accordingly in order to have his or her needs meet. The interface and dynamic interactions of the Transparent Charging Station provide the user with the tools to relate to the charging station and its comprising network; therefore, the user is able to actively navigate how to make the most out of his or her relationship with that network.

Thingformation attempts to satisfy a similar objective through its pictorial label. In clear and simple terms as a packaging label, Thingformation indicates what the landscape of the network behind the product is. The intention behind this label is to communicate to customers the complexities that lie behind a product before it is purchased, so that it can inform their purchase. It becomes the responsibility of the consumer to develop a sensitivity to that landscape and to decide how they would like to navigate or participate in it. There is a learning curve that comes with this label—some learning and experimentation with what the reality behind what these symbols represent. But this is a form of active engagement that supports Thingformation being transformed into something relatable. People have to develop a sensibility about the following: What is a good encryption level; is one level acceptable for some forms of data over others; what does trust look like; how does this compare to another product; what are the national or international bodies of data laws that I feel comfortable with? Thingformation provides people with the foundations upon which product owners can build their sensitivity to networks.

This is not unlike how young people who do their own laundry for the first time learn how heat or turbulence in a wash cycle may impact their clothing. There is a period of experimentation and failure, like any learning process, but then we learn how this corresponds to the wash labels. These symbolic forms of information are given meaning when combined with the sensibilities that we have developed. Taken together, we can then determine how we will engage with whatever networks we encounter. It is encouraging that these wash labels are reportedly consulted by four out of five customers to inform clothing purchases, a signifier that it is effective at helping people determine how they would like to relate to the product (American Cleaning Institute 2017).

Admittedly, the lack of immediate feedback and direct engagement with the network in the case of Thingformation is less ideal compared to the Transparent Charging Station. However, the success of Thingformation's processors (wash labels) is encouraging. Both Thingformation and the Transparent Charging Station provide context about the networks that they operate within and also the opportunity for users to realize its meaning and act accordingly. This is in contrast to merely being the benefactor or consumer of the output of these technologies, the networks or *ends*. Here through design, users are framed as engaged agents that are nodes within that network.

Discussion

Networks are complex and are becoming increasingly commonplace among the technologies that users engage with daily. Yet despite this prevalence, what is lacking are attempts to reveal the existence of those networks and efforts to make those networks relatable to the layperson. The urgency behind these efforts is mounting as these data-intensive and connected technologies become more sophisticated in gathering intimate data about users and potentially broadcasting it to others outside of the user's awareness and with significant impact. These networks are designed in such a way to do their work invisibly (see Wiltse this volume). Behavior is tracked surreptitiously, and that data is filtering into some other services, sometimes in unexpected ways. The question that this chapter considers is how design can offer opportunities to reveal the existence of these networks and our roles in them as users, and to contribute to making them relatable to the layperson.

Borgmann's critique of our relations and engagements with technologies provides a critical framing to approach to reflect on the relations between people and complex technologies (Borgmann 1984). Borgmann's writing preceded the widespread prevalence of networked technologies, and yet it is still a constructive and relevant foundation to examine them. While he opens many critiques of contemporary technologies and identifies the problematic conditions under which they flourish, there are not many insights offered as to how to correct these issues to make the technology more relatable. Furthermore, the conventions of product design practice today perpetuate the problematic relations that Borgmann had originally critiqued.

This chapter draws on conceptual design artifacts to provoke these conventions that give rise to the device paradigm and challenge the ability of users to relate to these technologies. Specifically, this research attempts to unmask networks behind the edifice of an artifact to understand it for its material and interactional properties. This research highlights the reality that networked technologies are defined by the relations that exist among different nodes, and more importantly that people are also nodes in these networks, playing a role in defining the purpose to those other nodes. At a fundamental level, networks become a material of the technology as they are part of what constitute the object and make it possible for it to perform its function. As a material, networks have qualities that shape our interactions with the technologies and also the shape of the output itself. Therefore, design needs to surface the materiality of these networks in order to first reveal their existence and to further contribute toward supporting their relatability. These conceptual designs illustrate how a product label and dynamic interactions and interfaces can contribute to this objective. In both cases, networks are materialized as something that requires the active engagement of people. The design of these technologies transforms networks into something that is accessible to the layperson and actively requires their engagement, thus facilitating relatability.

In revealing these networks and promoting their relatability, opportunities for networks and technologies to benefit from human input, engagement, and creativity are being opened up. Arguments are made in favor of reading and coding literacy as being a critical component for building a stronger society with more participation. The beauty of when information takes a conventional and relatable form is that that information is now positioned to benefit from the massive resources of human creativity. Language is constantly being adapted, developed, and expanded through people's active engagement with it. With more people able to read and write, members of society are able to record, share, reflect, transform, and delve deeper into the human experience. By opening up opportunities for networks to become relatable to laypeople, opportunities are also being created to empower people to contribute to society.

While reading the news, we can see a discomfort growing with the lack of relatability behind how networked technologies shape our society. How do news items spread through digital networks and become credible, how do filter bubbles shape our realities, or how does our personal data on platforms become a resource for others (Eslami et al. 2015; Hern 2017; Halpern 2018; Leetaru 2018; Meyer 2018)? There is a growing demand to reveal and make networks relatable, and it's time for the design of these technologies to facilitate that.

References

Afdeling Buitengewone Zaken, beyond.io, FROLIC Studio, The Incredible Machine, and Holly Robbins. 2015. "The IoT Manifesto." *IoT Manifesto*. May. http://iotmanifesto.org.
American Cleaning Institute. 2017. "Clothing Care Labels May Now Use Symbols Instead of Words." *American Cleaning Institute*. http://www.cleaninginstitute.org/clean_living/guide_to_garment_care_symbols.aspx (Retrieved May 30).

Amsterdam Institute for Advanced Metropolitan Solutions. 2017. "AMS Science for the City #5—Democracy by Design—22 Nov." *Ams.* November 20. http://www.ams-institute.org/news/ams-science-for-the-city-5-democracy-by-design-22-nov/.

beyond.io. 2017. "Beyond.Io." *Beyond.Io.* https://www.beyond.io. (Retrieved November 30).

Bihr, Peter. 2017. "A Trustmark for IoT." 1st ed. ThingsCon. https://github.com/openiotstudio/general/raw/master/publications/a_trustmark_for_IoT_thingscon_report.pdf.

Borgmann, Albert. 1984. *Technology and the Character of Contemporary Life.* Chicago: University of Chicago Press.

Dourish, Paul. 2014. "NoSQL: The Shifting Materialities of Database Technology." *Computational Culture* 4.

Dourish, Paul and Melissa Mazmanian. 2011. "Media as Material: Information Representations as Material Foundations for Organizational Practice." Corfu, Greece: Third International Symposium on Process Organization Studies.

Eslami, Motahhare, Aimee Rickman, Kristen Vaccaro, Amirhossein Aleyasen, Andy Vuong, Karrie Karahalios, Kevin Hamilton, and Christian Sandvig. 2015. "'I Always Assumed That I Wasn't Really That Close to [Her]': Reasoning About Invisible Algorithms in News Feeds." In 153–162. New York: ACM. doi: 10.1145/2702123.2702556.

Fallman, Daniel. 2009. "A Different Way of Seeing: Albert Borgmann's Philosophy of Technology and Human-Computer Interaction." *Ai & Society* 25 (1). Springer-Verlag: 53–60. doi: 10.1007/s00146-009-0234-1.

Fallman, Daniel. 2011. "The New Good: Exploring the Potential of Philosophy of Technology to Contribute to Human-Computer Interaction." In 1051–1060. New York: ACM. doi: 10.1145/1978942.1979099.

Giaccardi, Elisa, Chris Speed, Nazli Cila, and Melissa L. Caldwell. 2016. "Things as Co-Ethnographers: Implications of a Thing Perspective for Design and Anthropology." In *Design Anthropological Futures*, edited by Rachel Charlotte Smith, Kasper Tang Vangkilde, Mette Gislev Kjærsgaard, Ton Otto, Joachim Halse, and Thomas Binder, 235–248. Bloomsbury Publishing.

Halpern, Sue. 2018. "Cambridge Analytica, Facebook, and the Revelations of Open Secrets." August 21. https://www.newyorker.com/news/news-desk/cambridge-analytica-facebook-and-the-revelations-of-open-secrets.

Hern, Alex. 2017. "How Social Media Filter Bubbles and Algorithms Influence the Election." *The Guardian*, May 22. https://www.theguardian.com/technology/2017/may/22/social-media-election-facebook-filter-bubbles.

The Incredible Machine. 2018. "The Incredible Machine." Rotterdam. https://the-incredible-machine.com (Retrieved March 22).

Koskinen, Ilpo, John Zimmerman, Thomas Binder, Johan Redström, and Stephan Wensveen. 2011. *Design Research Through Practice.* Elsevier.

Leetaru, Kalev. 2018. "Why 2017 Was the Year of the Filter Bubble?." *Forbes*, December 18. https://www.forbes.com/sites/kalevleetaru/2017/12/18/why-was-2017-the-year-of-the-filter-bubble/#35563626746b.

Meyer, Robinson. 2018. "The Grim Conclusions of the Largest-Ever Study of Fake News." *Science*, March 8. https://www.theatlantic.com/technology/archive/2018/03/largest-study-ever-fake-news-mit-twitter/555104/.

Nansen, Bjorn, Luke van Ryn, Frank Vetere, Toni Robertson, Margot Brereton, and Paul Dourish. 2014. "An Internet of Social Things." In 87–96. New York: ACM Press. doi: 10.1145/2686612.2686624.

Orlikowski, Wanda J. 2007. "Sociomaterial Practices: Exploring Technology at Work." *Organization Studies* 28 (9): 1435–1448. London: Sage. doi: 10.1177/0170840607081138.

Robbins, Holly. 2018. "Materializing Technologies: Surfacing Focal Things and Practices with Design." PhD Dissertation. Delft, Netherlands: Delft University of Technology.

Robbins, Holly, Just Things Foundation. 2017. "The Path for Transparency for IoT Technologies." Edited by ThingsCon. *Medium*. June 12. https://medium.com/the-state-of-responsible-internet-of-things-iot/hollyrobbins-6d2f81512242.

Robbins, Holly, Elisa Giaccardi, and Elvin Karana. 2016. *Traces as an Approach to Design for Focal Things and Practices. The 9th Nordic Conference*. Gothenburg: ACM Press. doi: 10.1145/2971485.2971538.

Rosner, Daniela, Jean-François Blanchette, Leah Buechley, Paul Dourish, and Melissa Mazmanian. 2012. "From Materials to Materiality: Connecting Practice and Theory in HCI." In 2787–2790. New York: ACM Press. doi: 10.1145/2212776.2212721.

Wakkary, Ron, Doenja Oogjes, Sabrina Hauser, Henry Lin, Cheng Cao, Leo Ma, and Tijs Duel. 2017. *Morse Things: A Design Inquiry into the Gap Between Things and Us. The 2017 Conference*. New York: ACM Press. doi: 10.1145/3064663.3064734.

Wiberg, Mikael and Erica Robles. 2010. "Computational Compositions." *International Journal of Design* 4 (2): 65–76.

14 REFLECTION AND COMMENTARY

Erik Stolterman

Does there need to be a book in which scholars reflect on what defines a "thing"? It's a concept that probably sounds strange to most people, and with good reason. After all, there's nothing strange about things. We live with things, they are all around us, and humans have always lived with things. So, are the ways most people think about things not the "right" way, or what is the problem? The authors of this book make the case that it is important how we think about "things" and that it matters how we understand them.

For this to make sense, we have to leave the realm of everyday thinking. That is not what this book is about. Instead, we are dealing with scholarly or philosophical thinking. In the introduction, Wiltse makes it clear that the purpose of this book is an attempt to "ask fundamental questions about the role of things in human affairs and also how they might be designed differently to serve more desirable forms of life." In other words, we're not dealing with the usual way of thinking about things. The authors of this book are instead involved in serious and fundamental activities aimed at revealing and exploring different ways of thinking about "things." It is a philosophical and conceptual adventure, and, as such, it takes on a certain shape and form.

We should expect the adventure to lead us through unknown areas; explore new phenomena; open new questions; and, as a true adventure, challenge us in some way. It may be both difficult and uncomfortable to follow the authors through their reasoning and examples. But this is how new ground is explored. The ground may be rough in the beginning and overgrown without clear paths. It may even hurt when you try to go through certain areas. Slowly, by the attempts of many individuals who are creating paths and openings, we will begin to see the landscape and may be able to navigate it. And as most conceptual adventures of this type, this book does not offer simple and straightforward answers or any comprehensive solutions. The contributing authors do not provide definitive definitions or final frameworks or any kind of practical solutions. Instead, they all contribute by adding their own perspective related to the basic question that they all were asked to reflect on: *How to relate to these things that relate to us?*

Many of the authors start their exploration by stating why this is an important question and how they see their own contribution relating to the question. With such a broad overall question, it is not surprising that the book contains a wide range of reflections which are diverse when it comes to topic, perspective, and, not the least important, style. The reflections are nicely summarized and presented in the "Overview of the Book" section in the introduction chapter, so I will not engage in any summary of the book here. Instead, I will reflect on some aspects that have emerged when reading the book. I will also comment on what it is that we might learn from reflections of the individual authors. Ultimately, like the authors, I will not provide any comprehensive view on the topic. Instead, I'll join many of the authors and end with more questions.

It is fascinating to see a group of researchers from such diverse fields come together in a study of a common (everyday) phenomenon. It does not happen often. In most cases, the situation must be designed. Someone must decide what to study and how to study it and then invite scholars to be part of that investigative adventure. That's how this book came about. In this case, the phenomenon in question is something as simple and everyday as a "thing."

How is it possible that all these scholars are willing to discuss something that, to most people, is obvious and not even worth examining? What is the problem that intrigues them? A "thing" is, to most people, just that: a thing. Things have always been part of our lives and will always be. Why do we have to engage in a fairly complicated scholarly investigation of something that is obvious? Isn't this just another example of academics engaging in obscure deliberations for their own sake? Well, we can't deny that such deliberations are what many scholars love to engage in, but it is also the case that the scholars of this book are convinced that "things" are not as simple as they once were and, in fact, might never have been. It might be the most important insight from this book: There is nothing trivial about things.

Most humans live in environments that experience a growing number and diversity of traditional things. But we are also experiencing a radical shift in our everyday environments as a consequence of the infusion of computational abilities in everyday objects. The increasing level of "agency" or intelligence in artifacts and systems is stunning, and we can quite safely predict that this development will continue even more radically in the years to come.

Humans today are, in many cases, living in environments that reactively or proactively reason and act based on what is going on in the environment. Self-driving cars are a prototypical example. A car is traditionally a thing that was understood as a tool, that is, something that is under the complete control of a human. Now, cars are becoming actants, which interact with other vehicles, humans, and the environment in its full complexity and richness. To think about such a car as a tool in the traditional sense has become quite problematic and, in many cases, does not make sense. So, what is that car? Is it a thing? How should we understand it, and how should we talk about it? And what does it do to us? And how should we relate to it?

Similar developments are evolving in almost every aspect of human life. The sections of this book illustrate this quite well. The first section of the book, for instance, deals with care. Things are changing what care means. We do not only take care of things; things take care of us. Similarly, other sections investigate how these new things change *learning*, *controlling*, and *revealing*.

But the authors do not only address these new forms of things. Several of the authors discuss simple things that are not in any way technologically advanced or enhanced. This tells us that when these researchers investigate our relationship with things, they are not only responding to a reality being changed through technology but trying to explore any human relationship with things in a serious way. And maybe the challenge when it comes to understanding things is not related to the most advanced and intelligent things but to ordinary things.

Maybe the most profound revelations about the nature of things emerge when we examine the things we have always surrounded ourselves with, things we do not pay attention to in our daily lives. Some authors explore how these ordinary things relate to us and how sometimes, in order to understand them, we have to approach them as if they have real agency even though it is hidden under layers of everyday habits. The things examined in the book can be understood along a dimension, stretching from the simplest "dumb" object to the most advanced intelligent "being." But, no matter how they are defined or where they belong on that dimension, the authors agree that, as humans, we have to take a closer look at how we relate to these things and how they relate to us.

It seems obvious that contemporary things, defined as things that are enhanced by computational abilities and connected through vast networks, constitute a particular form of things and that there is a need for new ways to approach and understand them. The authors also make the case that if we do not engage in a deeper examination of things, we may end up in a reality where things take on roles and perform actions that we might find surprising, offensive, and, in many cases, less desirable or even dangerous. The realization that "things" are more than trivial seems to have stimulated these authors to engage in a range of sophisticated explorations of things of all kinds. Even if the authors do not present a common understanding, the realization that things are no longer what they used to be is another major contribution of the book.

Reading the chapters, it seems as if there is some consensus that "common sense" thinking about things is not sufficient when it comes to things and, especially, new forms of things. Each chapter offers its own alternative, some based on empirical analysis of a certain type of thing, others on philosophical examinations of fundamental assumptions about things. These alternatives are sometimes complementary but also, in some cases, contradictory. This is, of course, not a problem in the case of this book. The authors are not trying to compose the best possible understanding of things; instead, they try to ask new questions, to show where there are gaps, and to suggest possible ways forward. I will briefly mention some of them here.

It seems as if there are two ways of understanding "things": *things in themselves* or as *objects that mediate*. Even though the difference between the two is not easy to establish, they do provide quite different starting points and perspectives when it comes to investigating things. If we see things as objects in themselves, then they have their own agency and purpose, maybe even identity of some kind. If we instead see them as mediating objects, it means that we ascribe agency and purpose to something, or usually "someone," outside the thing itself. Someone is ascribing the purpose and the possibility to act to the thing.

This has consequences when it comes to how humans will handle "things." For instance, it makes a difference when we deal with other people and whether we believe they act based on their own beliefs and reasoning or if they act as a "means" for someone else's intentions. To most people, it is probably easier to think about things as mediating someone's intentions. A robot is only doing what its designer and programmer has told it to do. The same goes for a dish washer or a car. To understand things then becomes a tracing exercise, to trace its behavior back to whoever designed the internal mechanisms of the thing. To understand things in themselves is usually more problematic. It means that we have to see them as having their own intention and purpose, maybe even a mind or soul of some kind. Of course, in everyday life, people do this all the time. They see things as beings, as having some form of agency. We blame things, we kick them, and we scream at them when they behave "badly." But even in those situations, few would agree that their treatment of the things is a sign of the things having agency and intention by themselves.

The question is, of course, not only if things actually have agency. For some authors, the question is better framed as if it is practically better to think about things as if they have agency and allow the ontological question be irrelevant. This debate is not only philosophical. In the case of self-driving cars, this issue is highly relevant from a practical perspective. For instance, if a car behaves badly, should we punish the car, the driver, the designer, or something else? It may sound silly to punish the car, but if we take the perspective of things in themselves having agency, maybe that is the right thing to do. Some might see this as some form of evolutionary approach. We punish bad things, so they become extinct while good things survive. And we do this without punishing those who created the thing. Even though this approach works in nature, it may be few who would subscribe to it when it comes to things, since the cost in terms of failures could become high. However, if we, as some authors do, look at very simple things, maybe this is an approach that would make sense.

Inscribing intentions and behavior to "things" is what designers do in their attempt to make things function well and achieve their purposes. Many studies in science and technology have shown that designed things, apart from being able to perform an intended function, also and inevitably express some form of values. In many cases, designers are not aware of their own biases and the values that influence their designs, but sometimes this is done with the idea that the designed "thing" also should reflect certain ideals or values. Even simple things relate to us in ways that influence who we are and how we can navigate the world.

But, as some authors discuss, we are facing a new reality where designers are sometimes purposefully aspiring to inscribe values. This can be done with simple things, for instance, as with the example of sharp spikes in public spaces serving as a deterrent to people sitting or lying down. But what about digital things? Is the opportunity to inscribe values increasing or not, and what does that mean? As some authors discuss, we may not be satisfied with "things" that only reflect values. We might actually want them to have values and to act based on those values.

As Michelfelder discusses in her chapter, if humans care about "things," then "things" should care about humans. However, when we move in that direction, we also increase the complexity when it comes to ethical considerations in design. Who has the power to inscribe values, and what values are acceptable? Of course, since we already surround ourselves with things, this is already taking place. That is exactly what some of the authors point to. They argue that we are already living with things that have values. We have to pay more attention to how they relate to us and in what ways these things exert ethical "actions." Again, the argument is that things are never simple; they *do things to us*. Most people would agree that a robot will do things that impact us, but the authors argue that even the simplest of things do things to us, even though we may not be aware.

What if we take another step and not only relate to things but partner with them? What if things could be designed to be partners in human activities? Is that possible, and what would it mean? As some authors discuss, it could mean different things. First of all, to see things as partners does not have to imply that we see them as sentient beings that act intelligently and purposefully. Giaccardi proposes that we can include things as design partners in ways that would not require things to be anything more than what they already are, that is, just plain objects. Giaccardi argues that it would just require us to approach and "read" things in a different way, and when we do, things become partners in our examinations and explorations of the world around us. They can show us aspects of reality, they reveal issues, and they point to unseen and forgotten qualities.

We can, of course, push this question further, for instance, by asking what would happen if things could learn in the same manner as humans, and even more so? How would that change the way we understand things? Gransche explores this in his chapter on what might happen if things develop the ability to really know who we are. One consequence could be that humans rely more and more on things that can form systems which, over time, learn everything about humans—and maybe more than humans can know. Gransche explores such a development and warns us that it could lead to a situation where things are no longer only our partners but our superiors. As a consequence, humans might be locked into certain roles that only the system can control and will allow. The interesting aspect of this may be not a final and possible dystopian future—we've all seen that in many sci-fi movies—but the idea that it will be caused not by extraordinary intelligent robots but by large numbers of things operating in collaboration with the best intentions in mind when it comes to the well-being of humans. This would mean that we should not fear an apocalypse brought on by the singularity and killer robots but rather a wealth of comfortable designs that work together to establish systems and mechanisms that provide people with all they need in a way that drastically lowers their ambition to critically investigate their own reality.[1]

To me, this is another of the core contributions of this book. Living with technology and technological things is not all about the most advanced artificial intelligence developments but about the already-existing landscape of things that surround modern life. Radical change may not come through breakthrough technology but rather as a consequence of humans' inability to relate effectively or critically investigate their things' environment. The authors in this book clearly argue for the need to develop humans' ability to relate to such an environment, and, in their opinion, we need concepts and these types of investigations to give us some handles and tools to work with.

Another reflection that emerges during the reading is if there is a (weak but surprising) undercurrent of technological determinism throughout some of the chapters. It seems as if several authors are making the argument that "if we are not able to change our way of thinking about 'things' on a deeper level, technology development will lead to xxx." This is, of course, not technological determinism in its pride, but I do get a sense of almost urgency in some chapters. To me, this is an exciting observation. Is there an urgency? If so, why? If we see the transformation of things as a consequence of new technology, does it mean that the full potential of that shift will take place, no matter what we do? Or, can we, as researchers and designers of "things," shape the future? And the more focused question related to this book: Will a changed way of thinking about things actually influence the future of things? While thinking about the future, we may also ask how this transition or shift of understanding "things" relate to earlier shifts in our history of "things." Has there been similar shifts, and if so, did that lead to changed ways of thinking? It seems a historical perspective on how things have related to humans and vice versa could add some insights that forward-looking studies may miss.

At a meta level, we might ask other questions such as, can pushing new ways of thinking about "things" become too abstract, too theoretical, and maybe even counterintuitive? Of course, any new theoretical development goes through phases of some complexity before it can be condensed and simplified. The question we may ask after having read the chapters is if these ideas and theories can be "translated" into pedagogical (simpler and concrete) language that can be used in teaching and maybe even inform the public?

Would or could a new way of thinking change the way we live with things? Van Den Eede touches on these questions at the end of his chapter. He writes that a "new vocabulary does not become much effective until a critical mass of people start using it, or parts of it. And this is usually something that does not happen, or when it happens, it does so very slowly." It might be that transforming a way of thinking that challenges common sense and everyday language is a futile project. Van Den Eede notes, "How could ideas so far removed from common sense reasoning ever begin to transform it?"

To me, this becomes the overall challenge to all the authors and readers of this book. If we accept the basic premise and assumption in this book that there is a need for a new way of understanding "things" and we all continue to develop new terms and concepts that more appropriately capture the nature of these things, then what? How do we take this understanding and "language" and transform it into something that can resonate with people's everyday understanding of things and their way of seeing and talking about them? And maybe a final question: If we could do that, what kind of change would we see?

Note

1   This "mechanism" has been wonderfully described by Herbert Marcuse (1964), and I have discussed this in more detail in another text (Stolterman 2018).

References

Marcuse, H. 1964. *One Dimensional Man: Studies in the Ideology of Advanced Industrial Society*. Boston: Beacon Press.

Stolterman, Erik. 2018. "Herbert Marcuse and the 'One-Dimensional Man.'" In *Critical Theory and Interaction Design*, edited by J. Bardzell, S. Bardzell, and M. Blythe. Cambridge, MA: MIT Press.

INDEX

actor network theory (ANT) 139,
    148 nn.1–2, 167, 192, 199
  atom-networks 199
affective computing 34
Alexa 24–5, 32–3, 43–55
  caring of 48–50
  gendered setup 48, 54
  privacy issues 45–6, 50–2
  user reviews 47
  voice commands 49–52, 54–5
Amazon Echo 49–52, 61, 69, 78 n.19
  user reviews 46–7
Apple Siri 3
applied ethics 31–2
artificial intelligence (AI) 5, 34, 38, 44–5,
  129 n.24, 180, 185
artificial sensing 103, 105, 112–13,
  116–17, 124
assisted living. *See* Alexa
attachment-related objects
  ethical assessment 32–4
  existential experience 31–7, 39–40
  human-thing relationship *vs.* 35
  role of emotions 32, 37–8
  technical acting 62–3
  virtue ethics model 37–9
attachment theory 31–40
  insecure 36–7
  relating to things 35–7
augmented objects 174, 179–80, 182–3
augmented reality (AR) 9, 34, 163, 234
  classical post-phenomenology 174–6
  composite intentionality 174, 176–7,
    179, 184
  intentionality to relegation 179–84

behavioral change 15, 68, 90, 144, 151,
  154, 158

biases 126, 274
breakdown 1, 10, 181, 206, 246–8, 251, 257

care/caring
  circles 52–3
  digital assistants and 8, 24–7
  feminist ethics 19–20, 48
  virtues 38, 45, 49, 52–4
causal luck 94–5
CCTV (Closed-Circuit Television) 140, 142
cellphone 173, 180, 185 n.1
circumstantial luck 94
common sense 32, 39, 192–6, 205, 208–10,
  273, 277
constitutive luck 94–5
co-speculation 9, 216, 220, 222–4
CPS (Cyber-physical Systems) 64, 73

data analytics 103, 105, 112–13, 116. *See
  also* artificial sensing
data technologies
  design partnership 102–5
  thing perspectives 100–2
dataveillance 249
design partnership
  artificial sensing 112, 114–20, 124
  democratization of manufacturing 112
  DIY practice 112, 114–16
  emergent objects 108–12, 114–16, 122–3
  MakeDo community 114–16
  more-than-human sense making
    107–8, 113–14, 120–1, 124–6
  Resourceful Ageing 117–19, 122–3
  Taiwanese Smart Mobility 104–10
design research. *See also* Morse Things
    project
  definition 104
  empirical strategies 220–2

HCI and 234–5
human practice 127
industrial context 250–1
material speculation 221–2
networks perspectives 259–60
withdrawal, notion of 215–16
design theory 5–7. *See also* design partnership; design research
effects on future's openness 72–7
device paradigm 50, 256–9, 262, 266
digital assistants 8, 24–5. *See also* Alexa
do-it-yourself (DIY) 112, 114–16. *See also* MakeDo community

emotional learning 95–7
emotions 13, 32, 37–8, 53, 92, 158, 231
ethics
  digital objects 32–4
  interactive screens 154–7
  video games 84–92
  virtue 37–8
everyday life
  aspects 1–2
  interaction scenarios 69, 71
  people and things 120, 229–30, 251, 274
  scooters and Taiwanese 104–5, 107–9
  societal roles 5, 61
everydayness 32, 39–40, 179, 184
existential identity 89–92

fluid assemblages
  intentional relations 240–6
  multi-intentionality 240–1, 246
  need for breakdown 246–8

games. *See* video games
GPS 24, 105
Grand Theft Auto 84–5, 90

HCI (human-computer interaction) 5, 7, 216–18, 234–5
  things perspectives 218–19
home automation 25
hostile design. *See also* security cameras
  paradigmatic example 135–7
  philosophy of technology 138–40
  physical imposition 142–5
  post-phenomenological thought 138–9
  public spaces 135–40, 142–5, 147–8
  "script theory" perspectives 135–6, 139
  self-coercive 144

human actors 63–5, 67, 74, 76, 77 n.4, 143
human-computer interaction. *See* HCI
human learning 8, 45, 63, 68–9, 77–8 n.7, 78 n.8
humanness 217, 220, 224–5
human-technology relations 64–7
  autonomy/automation levels 64–7, 73–6, 77 nn.4–5
  classical post-phenomenology 174–6
  composite intentionality 176–9
  intentionality to relegation 179–80
  Latour's door example 181
  learning and caring 67–8
  machine learning 68–72
  multistability concept 138–40, 147
  types 64–7
  virtual identity (example) 180–4

identity 19, 67, 88–91, 95, 174, 180–2, 239, 274
inaccessibility 193
  consequences 201–4, 207
incompleteness 215–16, 220, 231, 233–4
intelligent system
  dealing and learning 61–2, 67–72
  pitfalls 73–7
  technological autonomy 64–7
intentional relations
  concealment by interfaces 245–6
  multi-instability 243–4
  multi-intentionality 240–1
  "reverse intentionality" 242–3
interaction design 65, 164, 225, 256, 258–9
interactive screens 151–2, 154, 156–68. *See also* Product Impact Tool
interfaces 16, 71, 87, 240, 244–6, 256, 258, 261–2, 264–5, 267
internet of things (IoT) 8–9, 23–6, 50, 64, 73, 216, 218, 221–2, 231, 233. *See also* privacy
intimate relationships 22–3, 27
iterative learning 93–4

Jibo 61, 69, 71–2

Kant, Immanuel 64, 93, 165, 167–8

learning. *See also* human learning; machine learning
  definition 68

INDEX 279

process 63, 69, 81, 265
systems 62–4, 66–7, 69–71

machine learning 45–6, 61, 63, 67, 77–8 n.7, 99, 102–3, 116–21, 193, 203, 243
　algorithms 102, 123–4
　structural changes 69
MakeDo community 114–16
materiality 100–1, 112, 123, 140, 163, 192, 257, 267
moral luck 92–5
Morse Things project 217–34
multiple identities 85, 88–92

network technologies
　interaction design 256, 258–60, 262, 265, 267
　labeling system 260–1
　materials, definition 259–60
　relatability concept 255–7, 260, 264–5
　Thingformation 260–1, 265–6
　Transparent Charging Station 260–6
normative autonomy 65–6, 74–6, 77 n.4

object-oriented ontology (OOO)
　betrayal of common sense 193–6
　Cartesian categories 191, 193, 195, 200–1
　classic philosophy of technology 196–8
　commonsensically thinking 191–3, 196, 204, 209–10
　inaccessibility 193, 199, 201–5, 207, 210
　material and immaterial concepts 199–201
　philosophies of 196–201
　uncertainty 193, 196, 201–4, 207–8
OBSERVE project 154, 156, 160, 162, 168
operative autonomy 65
ordinary things 3, 7–8, 31, 33, 36–7, 39–40, 273
　ontological approach 31–2
ordinary wisdom 37–8

Pepper 61, 70–2, 78 n.8
personally identifiable information (PII) 15
philosophy of technology 4–5, 7, 9
　and design 6, 234–5
　hostile design 138–40
　Morse Things 217–18

Pokémon Go 9, 34, 173, 176, 181, 184–5
postphenomenology. *See* technical mediation
privacy
　autonomy and 19–21
　care-theoretic model 8, 24
　concept 18–19
　consent and 22–3
　intimacy and 19, 21–5, 27
　legal-juridical model 18–19
Product Impact Tool 151–4, 157–63

Resourceful Ageing project 116
resultant luck 94
robots 3, 24–6, 34, 38, 43, 59, 61, 64–5, 67, 72, 139, 166, 226, 231, 274–6
Romeo 71–2, 75

security cameras 135, 138, 140–5, 147–8
　confederacy logic 143–5
　self-coercion 144–7
smart homes 25, 59, 66, 72. *See also* Jibo
smart mobility 104, 107–8, 124, 126
smartphones 2, 33–6, 45, 54
smart technologies 1–2, 10, 25–6, 33–6, 45, 51, 54
Spotify 70, 75, 240–1, 245, 250
strategic autonomy 65, 67
surveillance capitalism 2, 10, 70, 243, 248–51
surveillance technology 19, 135, 137, 139–47, 167, 185, 249
　multistability 138–40, 147
　public-space issues 135, 140–8

technical acting. *See also* intelligent system
　attachment-related objects 62–3
　goal choices or normative orientations 72–7
　handling and learning concepts 67–72
　personalization 70–3
　realization gap 63
technical mediation. *See also* interactive screens
　ambivalence 167–8
　awareness and responsibility 157
　dystopian technology 166–7
　impact and ethics 158–66
　legal compliance 157–8
　legal regulation 157

"technoethical" or "technomoral" virtues 37
theory of mind 25, 52–3
thing-centeredness 217, 220, 224, 226, 234
things
    artificial performances of 103, 126
    attachment to 32, 34, 40, 182
    caring for 6, 34, 38, 48, 53
    as design partners 102–3
    other 3, 27, 100, 105, 107, 189, 209–10, 225, 227–8, 234
    technical handling 62–3
    withdrawal of 217, 219–20, 222, 231
    and us 7, 48, 101, 125, 152, 190, 215, 217–18, 220–1, 224, 229–32
thing turn 4–5

uncertainty 193, 196, 201–4, 207–8

video games
    checkpoints or autosavings 92–3
    emotional learning 95–7
    ethics 84
    experiencing power and weakness 85–8
    fundamental characters 82–4
    identity types 88–92
    iterative aspects 93–5
    moral perceptions 84–92
    multiple identities 88–92
    safe learning environment 84–97
virtue ethics 37–8

withdrawal 125, 127 n.5, 159, 215, 217, 235, 246
    things perspectives 219–20, 231–2
world-builder games 85–6

Zen 33, 38